Practical Computer Network Security

For a complete listing of *The Artech House Telecommunications Library*
turn to the back of this book.

Practical Computer Network Security

Mike Hendry

Artech House
Boston • London

Library of Congress Cataloging-in-Publication Data
Hendry, Mike
    Practical computer network security/Mike Hendry.
      p. cm.
    Includes bibliographical references and index.
    ISBN 0-89006-801-1 (alk. paper)
    1. Computer networks—Security measures. I. Title.
TK5105.59.H46 1995   95-23939
005.8–dc20   CIP

British Library Cataloguing in Publication Data
Hendry, Mike
    Practical Computer Network Security
    I. Title
    005.8

ISBN 0-89006-801-1

© 1995 ARTECH HOUSE, INC.
685 Canton Street
Norwood, MA 02062

International Standard Book Number: 0-89006- 801-1
Library of Congress Catalog Card Number: 95-23939

10 9 8 7 6 5 4 3 2 1

Contents

Part I: Requirements and Risks

Introduction | **1**

1.1 ABOUT THIS BOOK

Security is one of those words we use without thinking; we know it is a good thing and it gives us a comfortable warm feeling. Even criminals like security systems, since they indicate there is something worth protecting. A million different products offer to "enhance the security" of your computer system, and they all do different things.

Just what do we mean by security and why is it so important in data communications? Is it affordable, or, indeed, achievable? What measures are appropriate to a "normal" business, one that is not involved in highly confidential work or million-dollar financial transactions?

This book sets out to answer some of these questions. It is designed to be read by the many people who have responsibility for the security of their systems, ranging from small businessowners who buy systems more or less off the shelf, programmers and analysts who have to implement measures appropriate to the type of business they are involved in, to senior managers who may have to delegate large parts of this work, but are nonetheless responsible for approving budgets and for the consequences if the security breaks down.

Because of this wide range, we have tried to avoid excessive use of computer jargon or complex mathematics. There are several more detailed books on encryption and software techniques for high-security applications, and these are referred to in the Bibliography, located at the back of this book.

In the rest of Part I, we look at the concept of security in data communications, trying to identify the many different things we can mean when we use the word *security*. Chapter 2 is about identifying the problem: a solution is always much closer when the problem has been defined. We look at the many dimensions of security and risk and at the different effects a failure can have on the business. In Chapter 3 we consider the business framework within which our computer systems have to work; often the computer network is not the weakest link in the chain. Sometimes it may be impossible to provide adequate security for the computer network because of failings in physical security or staffing policy.

Chapter 4 returns to look in more detail at the categories of risk faced in communications systems, while Chapter 5 considers where the weak points in the system may be, looking in turn at the hardware, software, operating systems, and network design. In Chapter 6 we describe some risks and danger points specific to a particular industry or application.

In Part II we study the technology available to combat these risks. In Chapter 7, we consider what types of technology are commercially available and aimed at the security market, and how these mostly cryptographic or password-related tools fit with the best available cryptographic or information theory. Are we making the best use of current science? In Chapter 8 we deal with the specific problems of managing cryptographic keys.

Chapters 9 and 10 look at the hardware and software tools available to most users, including some specific examples of proprietary products in the market. In Chapter 11 we stress the importance of a well-designed and well-managed scheme of access rights as a precondition for any realistic security scheme. We look at the design of access control systems in the context of the corporate structure and management style.

An important factor in determining the level of security that can be achieved in a network is the design and architecture of the network itself, so in Chapter 12 we discuss the specific problems and measures necessary in many common network environments.

We finish up in Part III by looking at specific sectors and application areas. Chapter 13 deals with common commercial applications of data communications, including intracompany communications, field sales systems, and electronic data interchange (EDI). In Chapter 14 we look at the special problems inherent in banking and financial applications, both for banks themselves and for their commercial customers. Another particularly tricky area is subscription services, where a very wide group of users must be given access; these services are dealt with in Chapter 15. Chapter 16 looks at other special applications, such as software distribution, and Chapter 17 contains a summary of the book's main findings and some conclusions.

1.2 CRIME, BLUNDERS, GLITCHES, AND CRASHES

In defining the scope of a book about communications security, it is important to understand the scope of security we are considering. Here we are trying to be as broad as possible in our scope: data communications failures or breaches of security may have human or electronic causes. This book is concerned with both of these.

We can break these down further:

- Human-caused failures may be either accidental or deliberate: although at this stage we do not want to go into motives, let us call these *blunders* and *crimes*, respectively.
- Electronic failures may be short-lived (they recover on their own) or more permanent (requiring engineer or software intervention). Using our shorthand again, we can call these *glitches* and *crashes*, respectively.

This book is concerned with all of these. Each one can have a serious effect on a business, depending on the way it manifests itself. Staff concerned with security tend to treat blunders and glitches as very much less serious than crime or crashes, to the extent that many methods used to deter crime sometimes increase greatly the likelihood of a serious blunder.

1.3 AIMS AND REQUIREMENTS

A data communications system is not an end in itself. It is a part of a more widely defined system which we can presume has a functional goal: for example, it may allow customers to place orders, salesmen to retrieve their messages, or banks to transfer money from one account to another. In fact, most data communications systems are shared by several such applications, and this situation is often where the problems begin. Systems may be shared by several different organizations for the same function, or by one organization for many functions. The public data networks (which are most often run by national PTTs (post, telephone, and telegraph) alongside the public telephone networks) are the prime example of this: they have to provide a service to many small users exchanging a few transactions a day with other small users or central services at the same time as linking the offices of multinational corporations who are in constant touch, transferring huge files at high speed.

These networks have to provide easy access for a very wide range of users at relatively low cost. These criteria conflict directly with the needs of security; few users would consider that the public networks for data require very high security, but commercial considerations mean that in practice the requirements for availability and reliability are very high and those for privacy and accuracy only a little behind.

There are many parties involved in a data communications system apart from the originator and receiver (see Figure 1.1). These parties may have different interests in the security of the system, but they usually do not have contractual relationships to enforce their requirements on the others.

In order for each party to ensure that its requirements for security are met,

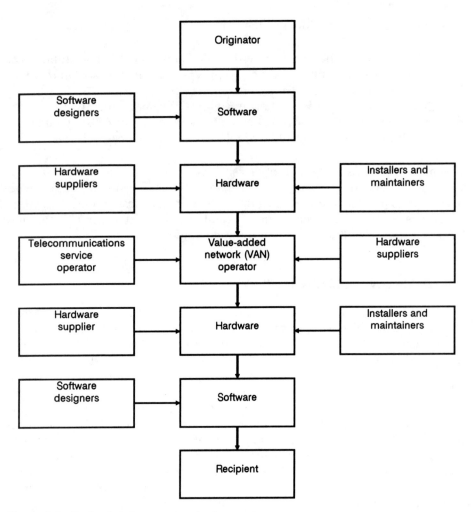

Figure 1.1　Parties in a data communications system.

it has to find some way of defining what those requirements are, and then building them into a requirements specification, service level agreement, or other contractual arrangement. The rest of Part I is about defining these requirements.

1.4　DEFINITIONS

Most readers will be familiar with the common computing terms used in this book, and we have promised to use as little jargon as possible. But some explanation of the specific terms used in the data communications literature, particularly those relating to security, may be useful. Abbreviations are explained in the text and in the Glossary.

Cryptology, Encryption, and Decryption

Cryptology is the science of codes and ciphers. Methods of keeping data secret tend to rely heavily on these. To keep data secret, we encrypt it using a key, and the data cannot then be understood by anyone reading it. The receiving party uses either the same or a different key to decrypt it (return it to the original data).

DES, RSA, and Symmetrical/Asymmetrical Keys

DES (data encryption standard) and RSA (Rivest-Shamir-Adleman) are the two data encryption standards most commonly used in modern data communication systems. DES was developed in the 1960s by IBM for the U.S. Department of Defense, and was subsequently published as a standard. It is a symmetrical key system; that is, the same key is used to encrypt and decipher the message. It is sometimes referred to as DEA (data encryption algorithm) or by its American National Standards Organization (ANSI) number, ANSI X3.92.

The RSA method was designed by a group of mathematicians who believed that it would not be possible to devise a code that could be deciphered using a public key without giving away the encryption key, and then proved themselves wrong! As the encryption and decryption keys are different, RSA is an asymmetrical algorithm. The relative merits of the two methods are discussed in more detail in Chapter 7.

Hashing, Scrambling, and One-Way Functions

Data are often hashed or scrambled before being encrypted. Scrambling involves cutting up the data and rearranging them according to a predetermined algorithm. Hashing is a more specific term, used for algorithms that reduce the randomness of a set of numbers and compress them in the process. A one-way function is an operation that always yields the same result with the same input, but it would be impossible to reconstruct the input from the output.

Serial, Parallel, and Synchronous/Asynchronous

These are all terms that describe how data are sent to or from a computer. They may be sent serially: along one pair of wires, with every "bit" of data following another, or in parallel, where all the bits of one word are sent at the same time, followed by the next word. In synchronous communications, everything is kept strictly in time by clock pulses, whereas in asynchronous communications, variable gaps are allowed between characters. You may also meet terms like 3270 and 5250, which are synchronous protocols used by IBM for communicating with terminals, and RS232, which is the most common standard for serial asynchronous connections.

Parity and Cyclic Redundancy Checks

A parity bit is a single bit added to a unit of data to maintain a fixed pattern. It is virtually always used in the context of the eighth bit in a byte, which can be set so that the sum of all eight bits is odd or even. This form of error detection is now little used, since most data transmission services have to be able to transmit eight-bit data bytes.

A cyclic redundancy check (CRC) is more useful; long CRCs can be very effective end-to-end checks for a communications system. They perform a simple operation (typically dividing by a polynomial) on all the data in a block or file, and the CRC is added to the block in the form of the remainder.

Network, LAN, WAN, VAN, and Node

A network, in this context, is a group of computers linked together. It may sometimes include terminals linked to the network by a terminal server. The term *local-area network* (LAN) is used when the network is completely contained within a limited area—usually a user's site or building—whereas *wide-area network* (WAN) describes a network using telecommunications lines or an external network operator to provide the connection between two or more sites.

A value-added network (VAN) is a WAN, usually operated by a service company for several customers, which provides more than simple connection between computers. The added services may include local call access from any point in the network, "store and forward" services, central databases, or simply a high level of security and availability guarantees.

A node is a point where the network is connected to a device (usually a computer, switch, or telephone exchange). For a WAN, a node is the point where an external user can connect into the network.

Analog, Digital, Packet, and PAD

An analog signal can have any value within a range, whereas a digital signal can only be on or off. Digital signals are much less likely to be altered when they are passed between two devices, and this is one reason why they are used almost exclusively by computers.

Converting sounds into digital signals also makes for more accurate transmission—hence the compact disc and the move towards digital telephony. Digital networks do not require modems—the data are transmitted directly.

On shared systems such as national networks, one cable (or a fiber-optic or radio link) carries many signals simultaneously; this process is called *multiplexing*. For analog signals, this can be done by shifting them up the frequency range by different amounts. With digital signals, it is more efficient to divide the messages into *packets*, each with an address on the front, rather like the pages of a letter each being sent in its own envelope, one after the other.

This requires a packet assembler-disassembler (PAD), located where the data stream is connected to the network (usually at the network node or telephone exchange).

Aspects of Security 2

2.1 CAUSES AND EFFECTS

Without going into dictionary definitions, there are many usages of the word *security*, even within data communications.

In this context, we are less likely to be concerned with *personal safety*, although anyone responsible for a computer system also has to be aware of the safety angle. Computer cabling is a hazard in very many offices, while strains resulting from long periods of monitor or keyboard use have received considerable publicity recently.

Security for a computer system is most likely to refer to the chances of the system breaking down or losing data. For a data communications system, these dangers are still present, but what is often seen as the main problem is the possibility of an outsider gaining access to the system.

As we saw in Chapter 1, these two risks can be caused either by human failure or deliberate acts, or by temporary or longer lasting hardware or software failures. We called them blunders, crimes, glitches, and crashes, and commented on the fact that many computer users are very concerned about crime, and pay much less attention to crashes, blunders, and glitches, which can cause just as much damage to a business.

In reality, it is not so much the causes we need to analyze as the effects. One good starting point is to look at the data being transmitted and the operations being carried out as a result of receiving the data:

- How critical are the data to our business? What effect would a 1% error or loss of data have on our profits for the month? What value would the data have to a competitor? Or, as in personal banking, is there a confidence factor that depends on a very high percentage of transactions being carried out correctly?
- What effect would there be on other parts of the operation if we were unable

to carry out this operation? Or what if the data were delayed by a few minutes or days?

It is important to have some idea of the answers to these questions, which put the data communications security issue into perspective and help to ascertain the benefit we may expect in return for any investment in this area.

2.2   COMMUNICATIONS SECURITY CRITERIA

Cryptologists and others who work on the technical side of data communications security usually work according to a number of criteria. You can use these same criteria when working out the importance of data and operations, as suggested in the previous section.

2.2.1   Nondelivery

The nondelivery issue deals with the likelihood that the data may simply not be received. There may be a big difference between knowing that we have not received data and carrying on processing without knowing that some data may be missing.

Data may not be delivered because the whole network is down or even because of a few critical links. Or it can even be misdirected, so that it arrives at the wrong destination. In nearly all cases, these are network problems: if it is a VAN, then the misdirection is probably the responsibility of the VAN operator. Public network operators usually avoid accepting liability for nondelivery, and, of course, on an internal LAN the full responsibility lies with the LAN operator.

2.2.2   Accuracy

Here we are concerned with random errors in the data stream: the data received are different from those transmitted. Probably the most important factor for any network is that the data are transmitted accurately from one end to the other: a single-digit error in an amount, for example, could have very serious consequences.

In practice, as we will see, any network with the slightest pretensions to security can detect small or large errors in virtually every case, and can nearly always correct them, sometimes by asking for the problem part of the data to be retransmitted.

The consequences of an error passing through the network itself and reaching the application program depend greatly on the data content itself: a single error within a text string, for example, is usually obvious and, if the data are not going to be processed further, is of little significance. Applications that would be more seriously affected normally have further checks on the plausibility of the data.

2.2.3 Data Integrity

Data integrity refers to the possibility of *alteration* of the data. A situation in which the data have been altered in such a way that they are still plausible and pass the network's checking mechanisms represents a more serious threat. The implication is that at least two parts of the data have been changed: the target and the check data.

Network users must ask themselves the question: what would happen if someone did intercept my message and altered it? If the consequences of this are serious, then the answer is likely to lie in either encryption, where an altered message yields nonsense when decrypted, or in check digits, which cannot be altered by an intruder.

2.2.4 Confidentiality

Closely linked to alteration is the issue of confidentiality or privacy: What happens if someone (a competitor, for example) is able to read the data we are transmitting? In some cases, this may not matter. Many of a company's data are not confidential at all; indeed, companies spend a fortune trying to ensure that as many people as possible read their material. Even slightly more sensitive information may not be significant, provided that only a part of the information is transmitted at any one time. In this way it is unlikely to be intelligible to any but the most persistent.

There are a few situations where the very existence of traffic is enough, without needing to know the content. For example, in a military situation, substantial traffic from headquarters to a new location is likely to anticipate action.

It is always worth remembering the test for confidentiality: How much damage will this do to our organization if the privacy is breached?

Other words associated with privacy in networks are *leakage* and *eavesdropping*. The first tends to imply a malfunction of the network, while the second relates to deliberate action.

2.2.5 Impersonation and Repudiation

How can we be sure that this message came from the person or organization it appears to come from and that the person who sent the message is authorized to send it? What happens if the person claims not to have sent it? How do we know that the right person is receiving the message at the other end, and how can we prove (later in court, if need be) that he or she did receive it?

This is particularly important for contractual documents: orders and payment instructions, for example. But it is wise to consider what the effect on other transmissions would be if someone were to *impersonate* the sender or receiver of a message or a set of messages. Another word associated with network privacy is *masquerade*.

Modern cryptography has a range of tools for "signing" a message or an acknowledgment in a way that cannot be altered without detection.

2.2.6 Causes of Failure

If we look at the different effects, or modes, of failure, we see that some of them are more likely to be caused by human error or by deliberate action, while others are more likely to be caused by hardware, software, or related problems. In Table 2.1, we look at the links between causes and modes of failure.

Table 2.1
Causes and Modes of Failure

	Causes			
	Human		*Electronic*	
Effects	*Deliberate (Crime)*	*Accidental (Blunders)*	*Short-Term (Glitches)*	*Long-Term (Crashes)*
Nondelivery	√	√	√	√
Inaccuracy	√	√		
Alteration	√	√	?	
Privacy breach	√	√		?
Impersonation	√			
Repudiation	√			

2.3 REQUIREMENTS SPECIFICATION

Using the criteria in Section 2.2 to produce a specification, the aim must be to balance the seriousness of the effect against the chances of its happening. A simple way to express these criteria is to use the *mean time between incidents* (MTBI), which is actually the same as a probability: an MTBI of three years is a 33% probability of an incident in any year. But using the MTBI method makes it easier to relate to the effect on the business.

Before drawing up a requirement specification for network security, you must consider whether all the data can be treated in the same way, or whether some data are particularly sensitive. It helps to know in advance whether your security requirements will in practice be easy to meet or very expensive.

Table 2.2
Sample Requirements Specification

	Minimum MTBI (Years) for		
	Undetected Incidents	*Detected Incidents*	*Detected and Corrected Incidents*
Nondelivery	5	0.5	0.01
Inaccuracy	5	0.1	0.01
Alteration	100	5	5
Privacy breach	10	10	N/A*
Impersonation	100	5	N/A
Repudiation	100	100	N/A

*Not applicable.

If all the data are similar, then you can draw up a chart showing the MTBI requirement for each criterion, perhaps for undetected and detected incidents, as in Table 2.2. Computer personnel can use data rates and other measurements to convert these MTBI requirements into bit error rates or other technical criteria.

Once the requirements have been set out in these very broad terms, we can start to consider the detailed risks and ways of addressing them.

Preconditions 3

An organization's security, operations, and data must all be viewed together. Too often, those responsible for security in an organization limit their role to physical security, while computer staff have widely varying attitudes toward security and a surprisingly free hand in implementing security measures. The only way to achieve satisfactory security is for senior management to have a realistic idea of what is required and achievable, and to be constantly aware of the risks.

Data communications security is only one link in a chain, and it can itself be compromised by poor security elsewhere. Much of what is thought of as computer fraud may actually be poor management control systems. A recent high-profile example was the case of Kidder Peabody in New York, which allowed a trader to use a loophole in the company's accounting system to book $350 million of excess trading profits. Although not strictly a data communications failure, the control that would have prevented this, matching the component parts of the deal, is probably the most important management control mechanism in an EDI system, and it is often omitted.

Although several of the factors featured in this chapter will also be mentioned later, they are brought together here, right at the beginning, because this is where security begins. There is no point in designing a sophisticated encryption system if an intruder can walk in and unplug it.

3.1 SYSTEM DESIGN

3.1.1 Business Process Design

Data communications requirements tend to grow with the firm, with new needs being met as they arise. This is not necessarily bad: to design too far in advance can lead to inflexibility and overinvestment.

But solving each problem as it arises can also create problems: today's problem may be a result of yesterday's solution. A sequence of expedient solutions may make life very difficult, whereas there may exist a single solution that solves the whole issue. One company had an array of locks, management controls, and signing-in procedures to control access to the computer room, but still found that these were circumvented by senior management. The solution was to subcontract all computer services to another company, which ran them from its own site.

From time to time, it is necessary to reevaluate the whole business process and decide what activities are really necessary. By avoiding unnecessary activities, you avoid the security problems associated with them, while a shorter, simpler path makes errors less likely and makes it more difficult for a miscreant to hide his or her actions.

This idea has achieved the status of a management fad in the early 1990s, and is called *business process reengineering*. Like many of the creative thinking processes described in this book, it is usually better carried out by one or two top managers over a beer than by an army of consultants.

3.1.2 Computer System Integration

Another frequent source of errors, and of opportunities for some to hide their deliberate actions, is where information has to be transferred from one computer system to another. Rekeying almost always introduces errors. Within a company, it should normally be possible to have a single integrated system; even if different equipment has been bought at different times, an integrated design can and should be drawn up, showing the links between them.

Between companies, EDI provides the answer: commercial EDI systems can be made very secure, and there does not seem to be a single documented case of a significant security breach on a major EDI service.

3.1.3 Ergonomics

Good ergonomic design dramatically reduces human errors. Well-laid-out data capture pages make errors clearly visible, and good software checks on input can help operators rather than obstructing them.

Time-out periods must be carefully chosen; an operator who knows there is a limited time period for entering a set of codes may be tempted to rush or take shortcuts. On the other hand, vacant terminals must be logged off or put into a secure state as quickly as possible. A good compromise is usually to issue a warning tone after, say, a minute, and log the terminal off after a further minute. Security-minded systems may make the periods much shorter.

Physical layout of the workspace is an important security factor; operators should be visible without feeling that people are looking over their shoulders. As a general rule, operators at screens should face a wall, so that they have their backs into the room. Poor design may encourage operators to browse through parts of the system that they should not be viewing, or may allow others to access the computer system while seeming to leave a note on a desk.

3.1.4 Network Design and Selection

Transparency and simplicity are the watchwords here. Some of the most powerful networks are actually the easiest to understand, but too many companies have installed complicated networks that have many more features than they require. They thus leave themselves in the hands of the one person in the organization who claims to understand the network—and who can judge whether he or she does in fact understand it?

Although there is considerable merit in the argument that networks should be able to accommodate a little more than the foreseeable growth, complexity is the enemy of security. Transparency should be a specific aim of a network specification. While many networks can provide stunning performance on known tasks, they are also capable of incomprehensible blockages.

The network suppliers' reliance on buzzwords is part of the problem here: in reality, it should be no more necessary to know the protocol used or the amount of RAM provided within a computer system than to know the formulation of a detergent or the horsepower of the pump motor in the washing machine before doing the laundry.

3.1.5 Hardware and Software Design

Hardware designs following good current methodology will not introduce security problems. Some older designs—or designs following more relaxed rules, such as for electromagnetic interference—may actually make it easy to breach security by picking up the high-frequency radiation from a computer bus or from unshielded cables.

Careful design can even reduce or eliminate some problems. For example, it is best if equipment cannot be unplugged while it is switched on, and better still if the network can detect when a piece of equipment is switched on or plugged in, and vice versa. Simple features such as keyboard locks can be easy or difficult to use.

We already touched on software design in the discussion on ergonomics. A structured approach to design and software organization is also essential in a security-aware environment. A wide range of structured design and programming tools is now available.

3.2 PHYSICAL SECURITY

Another essential prerequisite for any good security system is access control. No amount of encryption or high-security design will help if the equipment itself is stolen, vandalized, or simply switched off so that the janitor can plug in a vacuum cleaner.

The easiest and probably the most common way to breach the security of a network is to "borrow" a password that someone has left lying around, has written down, or just enters in front of your eyes.

It is impossible for a book like this to say what physical security is necessary in any given case. Again, the threat must be evaluated against the criteria we have discussed, and the likelihood according to criteria we will come to later. The main aim of physical security should be to avoid providing temptation: few locks will resist the really determined burglar, and even a disenchanted employee can wait a long time for an opportunity to slip in unnoticed if he or she knows where a key piece of information is kept.

A general template for access control procedures is shown in Figure 3.1. This is by no means comprehensive, but does indicate some of the most important areas for consideration.

3.3   ORGANIZATIONAL INTEGRITY

Many organizations invite security problems through the types of personnel they recruit, the way employees are managed, and the example set by senior management. Many successful companies get where they are by an aggressive approach to their market, to competitors, and to competition within the company. The emphasis is on results, often with little regard for the methods that achieved those results. Hire-and-fire policies, in particular, encourage employees to damage the company if they know that their jobs are in danger.

The chief executive is sometimes a part of the problem: to gain an advantage on a competitor, there can be a very thin line between "breaking the rules" and breaking the law. Many employees see their chief executive as a crook (whether or not his activities are defined as criminal by the law), and they feel no qualms about following his example.

It is inconsistent for a company to rely solely on the letter of the contract in its dealings with its employees, and then expect trust and loyalty back. This inconsistency is seen by employees as unfairness and is liable to be rewarded with untrustworthy behavior.

Security must be a consideration at all levels of employment. Even staff who do not normally have access to critical information may acquire a position of trust, or may be given exceptional access because someone else is on holiday. References must always be taken up and verified; for senior appointments, it may be necessary to verify curriculum vitae information as well. A finance director in the United Kingdom who defrauded his company of several million pounds was found to have several previous offenses; but he had joined his new company in a relatively junior role, for which a full security check was thought unnec-

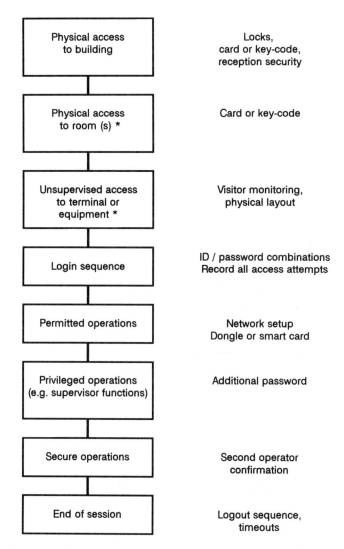

Figure 3.1 Procedures for access to a data communications system. *These must apply to all data communications equipment, not just user terminals. For example, telephone exchange equipment must be located in a secure area.

essary. As he rose up the ladder, it seemed unthinkable to carry out security checks on a person who had become a liked and respected member of the team.

The organizational structure is also important: there is always an informal structure in an organization, reflecting relative personalities rather than job definitions. The informal structure often translates into control of resources (not just the stock of pencils!) and may have a significant effect on security measures and computer system access.

The organizational structure can also create or minimize opportunities for collusion between members of staff. It is often wise to follow a strict policy of separating relatives and sweethearts, particularly so that there is never a situation where one checks or supervises the work of the other.

One of the best ways of minimizing deliberate attacks from within is to ensure that no one has sufficient motive to commit the attack. The possibility of financial gain, set against an atmosphere of trust, is rarely enough for people to plan to cheat their company (although they may submit to temptation if a good opportunity presents itself). Where the company, or its owners or managers, has in some way aggrieved the employee, the chances of a planned attack rise greatly.

3.4   CONTRACTS

When a part of a data communications system is subcontracted (as it nearly always is), the contract must reflect the security requirement and security specification. If, as in the case of a public telephone or data service, the contract is common to all users and cannot be altered, then the design must reflect the contract. There may be several options or service levels, in which case the choice of level should reflect the specification.

It is worth remembering the several aspects of security referred to in the previous chapter: contracts for data communications services frequently refer to availability, with no mention of other aspects.

Risk Management

4

Certain skills and procedures are applicable to managing all kinds of risk; data communications are only one subject to which they can be applied. In order to establish a policy and decide what countermeasures are appropriate in our case, we must start by categorizing the risks. This is done in various ways by looking at:

- The object that is threatened (a piece of hardware, some information, or a process or other equipment controlled by the data);
- The grade of risk (the damage that could be done in the worst case);
- The level of threat (the likelihood of a successful attack);
- The constraints within which we must work.

We then look at what countermeasures might be effective in each case, and what their cost is. Once these facts have been established, we can balance the risks with the countermeasures to produce an appropriate set of measures for our particular case.

4.1 OBJECT THREATENED

What is the threatened object? Is it a modem someone might want to steal? The information on how your new process works? The details of an order? Or is it possible that a wrong command being passed over the network might cause a valve to open and dump poisonous waste, or cause you to lose control of a satellite?

Most often, in a data communications system, the data are regarded as the object of the threat, whether accidental or deliberate. The information may have an intrinsic value: a mailing list, for example, can be sold on the open market. Or it may be fairly easy to attach a value to it: the value of an order, for example. However, in the majority of cases, the real threat is not the direct attack on the data, but how the attack affects some other process or business arrangement that depends on the data reaching the other end without being tampered with.

It is therefore important to ask the question: What is the worst thing that could happen if these data were lost, altered, or seen by someone else? Or if the other person claimed they had not sent or received it?

Perhaps the chance of the worst case happening is really very small indeed—so small that we really can ignore it. In that case, we should look at a more likely situation. This should, however, be a conscious decision based on a knowledge of the relative costs of the two threats.

4.2   GRADE OF RISK

Having ascertained what is threatened, we can then decide what grade or value of risk is involved. Is national security at risk? Is there a significant danger of death or injury? Could this threat endanger the whole business? Would we simply incur additional costs, which may be considered part of the cost of running a data communications system? Or would it just be a nuisance?

If possible, we should assign a money value to the threat. Where national security, life, or the survival of the business are at stake, it is often difficult to do this, but it is important to remember that there is no way of making any system 100% secure, and the cost rises very steeply as we try to approach this level.

4.3   LEVEL OF THREAT

Having listed the objects under threat and the grade of risk or value involved, we must now decide how likely it is that there will be a breach of security or an incident of some kind.

Is the object always vulnerable? This could apply, for example, to data passing over a telephone network, particularly an old-fashioned network or one where high winds often bring the lines down. In such a case, we know that an incident will happen from time to time.

Is it only vulnerable under some conditions? This might be at night or on weekends, for example, or over a shift changeover. Many systems are vulnerable when backup procedures or equipment are being tested, or while engineers have access to the system for maintenance or upgrades.

Is it susceptible to attack? In this case, a threat to the object would require some planning and perhaps collusion.

Do we really believe that the object is secure under all practical conditions? If so, we should not take any further steps unless the grade of risk is of the highest order, which implies that significant security measures have already been put in place. To say that our object is highly secure, we must be able to apply mathematical tests and show that we have considered all possible modes of failure. These proofs require a very high level of discipline and analysis and are likely to be beyond the level of most systems considered here.

In Chapter 2, we looked at a scheme for which we defined security requirements in terms of a minimum MTBI. We can classify the level of threat in a similar way: either as an MTBI or as the probability of a breach occurring at least once in a year. For more likely events, we can use incidents/year.

Multiplying the grade of risk by the probability gives us a measure of the likely cost of not protecting against this particular threat. Table 4.1, which you can compare with Table 2.2, gives an example of this.

Table 4.1
Risk and Likelihood

Event	Cost ($)	Probability Expressed as Incidents/yr	Probability × Cost
Telephone system fails <5 min	10	10	100
Telephone system fails <1 hr	1,000	0.1	100
Competitor sees client data	100,000	0.1	10,000
Client file is destroyed	500,000	0.05	25,000
Poisonous waste escapes	5,000,000	0.01	50,000

It is often much cheaper to reduce the likelihood of an event than to eliminate the possibility of it happening altogether. This is often the same as saying that you should look at the environmental factors before applying a technological solution. We mentioned earlier that the level of threat can often be reduced significantly by removing the temptation. Physical controls should ensure that people who might represent a threat do not even see the object they might threaten or the means of threatening it.

4.4 CONSTRAINTS

Having identified the threats and the possible cost of ignoring them, we are in a position to consider what countermeasures would be effective and affordable in each case. Most of Part II of this book will be concerned with the technological measures that can be taken to protect against the threats we have been discussing. But we must also be aware of the cost of these measures and other possible constraints.

4.4.1 Cost

Depending on the measures in question, there may be capital costs for the initial purchase of hardware, software, or accessories. Hardware also has an associated maintenance cost, while software may be subject to license fees. And there are likely to be operating costs and increased overheads associated with any security management system.

It can be difficult to justify high capital costs, particularly where there are existing installations and where strict capital budgeting schemes are in place. All too often, there was inadequate provision for security when the initial cost justification was made. Subsequent capital spending will yield no returns, although it may avoid losses—it is a form of insurance.

The insurance analogy often helps to justify ongoing operational costs, which are more likely to be within the control of a departmental manager. It is therefore often easier to gain approval for a moderate level of operational spending than for a capital expense, which would in the medium term save money.

4.4.2   Legal

Some security measures are illegal or prohibited by contract. Such restrictions apply not only to wiring up sensitive equipment to a high-voltage source, but also to many forms of encryption.

Governments in several countries, such as the United States, want to control all forms of encryption used so that the national security services will always be able to break the code. While this control may be useful in a tiny minority of cases, the harmful effects on true commercial security and the development of cost-effective methods can scarcely be imagined. In practice, if you can buy or develop an effective encryption scheme, you can use it. But you may need to obtain legal advice before selling or exporting it.

Some techniques may be excluded by contract. Many commercial data communication services, particularly VANs, restrict the security methods or encryption that can be used. This restriction is more reasonable, since security needs to be designed from end to end and balanced within a service. Some encryption techniques may interfere with network control. None of these restrictions is likely to apply to any high-security network, which will have been set up to accept blocks of data and encrypt them as such, without any regard for the context of the data.

4.4.3   Time

Time is a very powerful factor in providing and enforcing security. It can work for or against you, and the skillful designer will always ensure that it is on his or her side.

It takes time to crack a code, and if you can ensure that the code is no longer valid by the time it is cracked, then the information any would-be hacker has gained is no longer useful. For this reason, date and time fields are nearly always included in the raw data when message authentication checks (MACs) are being calculated. In some cases, time should be measured down to fractions of a second.

All forms of encryption take time. Another, very powerful security method is to use precisely calculated operation times within your algorithm so that it can only be decoded by hardware identical to the encoding hardware.

Finally, there may simply not be enough time to perform an ideal operation at the speed of your data transfer. This limitation applies particularly when the decoding has to be done in software, and is one reason why software in data communications encryption schemes is usually restricted to key management, with the actual decoding done in hardware. We will come back to this in Chapters 8 and 9, when we talk about the tools in more detail.

4.4.4 Technical

Very rarely, you may find that what you want to do is simply not feasible technically. It may impose excessive overheads, or the data rates involved may be beyond the capability of the technique. Even when a technique is feasible, however, it may still be too expensive in a given application.

Cryptology is a science that has fascinated many mathematicians over the years, and they have often been able to prove absolutely whether or not any given technique will work. The implementation of their theories in hardware or software, at a price the market will bear, is no less challenging a task.

4.4.5 Social

There remains the danger that the use of certain technologies, or indeed the protection of certain classes of data, may be regarded as antisocial or socially unacceptable. There are some strange examples of this. The French seem to regard it as unacceptable to scrutinize a signature, even on a check or other document that is legitimized solely by the signature. The British have a general dislike of identity cards, which makes them open to abuse. Fingerprint or thumbprint identification schemes are widely regarded with suspicion because of their traditional link with criminal investigation.

Many security schemes depend on a reliable method of identifying a person. Social attitudes inhibit the development of some of the potentially more reliable methods: retina scans and DNA tests in particular. These methods may, however, be developed in a purely military context.

In other cases, it is regarded as unacceptable to protect particular items of data: security is sometimes the opposite of trust. A widespread trend toward more openness in business leads employees and others to believe that any data needing to be concealed must be in some way underhanded or shameful.

4.5 BALANCING RISKS AND COUNTERMEASURES

When the risks have been listed and a value or grade attached to them, we can compare them against the cost of possible countermeasures.

There are several computer models that can be used to simulate protection measures; the risks, threats, and constraints can be fed in and parameters changed to yield the lowest-cost acceptable solution. More often, however, the risk management process must be performed manually, in which case a number of shortcuts may be useful.

Although in principle we should not try to protect the whole business, just those parts of it that may be at risk, it is often more cost-effective to view the security measures as a series of layers: outer layers have less security and the innermost layers have the highest security.

So we take the risks and group together the highest grade items. We decide the minimum level of security required and try to identify a method of providing that security within the value we have assigned to the risk. In practice, we will do this for all the layers together: the cost of providing low-level security for the whole system is usually higher than the cost of the ultimate security of the innermost layer, but it is essential for the inner layers that the outer layers are also protected.

In Figure 4.1, we give an example of the layers of protection that might be applied in a company with a computer system and external data communications. The authentication and encryption functions will be explained in Part II.

The key to this process is the need to balance the cost of the countermeasure

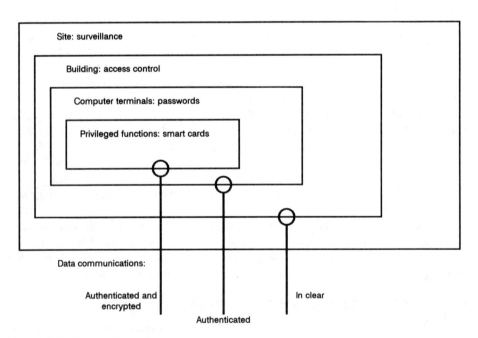

Figure 4.1   Layers of security.

against the risk. While no risk should ever be ignored, there are times when it is better to take a calculated risk and remain exposed to the threat than to spend a fortune protecting against it. Cost-versus-risk decisions should be explicit and their implications made clear.

4.6 STANDARDS

4.6.1 ITSEC and TCSEC

Standards have been developed that allow users to group their requirements into recognizable categories and to identify products that meet these requirements. The most important standards are:

- The Trusted Computer System Evaluation Criteria (TCSEC) [1,2], originally published in 1983 and administered by the National Computer Security Center (NCSC) on behalf of the U.S. Department of Defense.
- The European Information Technology (IT) Security Evaluation and Certification Scheme (ITSEC) [3], for which the national standards of France, Germany, the Netherlands, and the United Kingdom have been brought together since 1990. Published under the auspices of the European Commission.

The two standards are comparable but not identical, and it is also possible to make comparisons between them and the previous U.K. and German standards.
   Products are tested against a security objective. ITSEC defines three levels of security (basic, medium, and high) for each of eight generic functions:

- Identification and authentication;
- Access control;
- Accountability;
- Audit;
- Object reuse;
- Accuracy;
- Reliability of service;
- Data exchange.

The security objective consists of a detailed description of the functions performed by the product, under each of these headings, explaining if necessary why they will increase security. A product that resists casual attack has achieved basic security, but it may be defeated by a knowledgeable attacker. The medium level should withstand attack by anyone without considerable resources or time, while the highest level is reserved for products judged to be proof against all normal practical risks of attack.

The testing and evaluation schemes verify a product against its chosen criteria only. ITSEC accords a product one of six levels of confidence in its correctness (E1 to E6). For the lowest confidence level (E1), an informal (plain English) description of the security function and the way it is achieved is adequate, and testing is restricted to functional tests and documentation review. As the level of security demanded increases, so the requirements for proof and detail in the design increase; at the highest level (E6), formal descriptions of the architecture, design process, implementation, and testing are required in order to give an exceptionally high level of confidence that the product will meet its security objectives.

For ITSEC, almost any combination of security objectives may be chosen, and so it is important to know and understand these objectives in order to judge the suitability of the product for a particular purpose.

The U.S. TCSEC is structured somewhat differently, and its seven classes (from the lowest level D to the highest A1) incorporate both functionality and confidence requirements. Products meeting TCSEC requirements are therefore more likely to be homogeneous than their ITSEC equivalents.

4.6.2 Others

Other bodies concerned with computer network security include the International Standards Organization (ISO); the International Telecommunications Union (ITU), which has taken over the role of the CCITT; and ANSI.

ISO has published a document describing the Security Architecture for Open Systems Interconnection (OSI): ISO 7498.2 [4], a framework document that discusses the threats and appropriate security services within an OSI network.

More detailed recommendations for implementation of these services are given in ITU standards X.400 (Message Handling Systems), X.500 (Directory Systems), and other standards relating to OSI implementations. These standards are required reading for developers rather than users of these systems.

ANSI, for its part, has taken the initiative in standardizing several of the specific functions and algorithms required for data communications security. Of particular importance are the ANSI standards for the DEA (X3.92), MAC (X9.9), and PIN Management (X9.8), which have now been adopted by ISO as the standards in these areas.

Other bodies becoming involved in the definition of security standards include the Comité Européen de Normalisation (CEN) and the European Telecommunication Standards Institute (ETSI).

References

[1] "Trusted Computer System Evaluation Criteria" ("The Orange Book"), U.S. Department of Defense, 1985.

[2] "Computer Security Subsystem Interpretation of the TCSEC," U.S. National Computer Security Center, 1988.

[3] "Information Technology Security Evaluation Criteria (ITSEC): Provisional Harmonised Criteria," Commission of the European Communities, 1991.

[4] ISO 7498.2, Open Systems Interconnection Reference Model: Security Architecture.

Hardware, Software, and Networks | 5

Most of what we have discussed in the previous chapters applies to all types of risk. We now need to look in more detail at the elements of a computer network and the specific threats associated with them. It is worth asking why there should be so much emphasis on security in data communications. We do not send our mail in armored cars, so why do we need to take such strict precautions when we are sending data around?

The first answer is that a computer network is very complex; unless we pay attention to security, problems are guaranteed. The chances of an error or some unauthorized monitoring passing undetected are greater than in any manual system. The second reason is that, as we shall see in Part III, the scope of very large fraud or catastrophic errors is much greater with a computerized system than with a manual system. Computer-assisted fraud nearly always involves much larger sums of money than fraud committed without computer assistance.

The problems are aggravated by the barrier that still exists between computer people and the rest of the world. Many decisions that properly relate to the running of the business are left to computer staff whose knowledge of the rest of the business is at best imperfect. Or managers must make decisions concerning the selection of hardware, software, or encryption tools with only the sketchiest understanding of the relative merits of the available products. It remains to be seen whether a fresh, computer-literate generation will still suffer from these problems.

In this chapter, we will address some of the complexity issues and try to point to some of the vulnerable areas that may need more detailed investigation. It is important to understand the role of each element of the system in order to understand how it may be affected by a breach in security or how it can help security.

5.1 HARDWARE

The hardware consists of those parts of a computer system you can kick. If they can be kicked, they can also be stolen, disconnected, powered off, or tampered with. Wires may be connected to a piece of hardware to monitor electrical signals, and some hardware items (notably terminals and computer displays) even do the monitoring for an eavesdropper.

Theft. Items that are visible outside a building are particularly prone to being stolen: UHF and satellite antennas are popular targets. There is a ready market for stolen personal computers (PC), printers, and modems, although these must be recent models and preferably well-known brands. From a security point of view, rack-mounted industrial PCs are very much better than desk-mounted items—and they are usually more reliable and easier to upgrade, as well.

Power. Items are often powered off by mistake—many offices do not have enough power points and thus invite the cleaners to disconnect some critical piece of equipment in order to vacuum. At the minimum, every plug and socket must be clearly labeled. Communications equipment should really be powered from a separate switched spur; if a plug is used, it should be locked in place. Depending on the size of the building and local wiring regulations, it may be possible to put all the switches and protection at one central point (thus also allowing better power management). Or the switches may be central but with local circuit breakers close to the equipment. If a good network management system is in place, items losing power will be spotted quickly and the effects minimized. It is also possible to monitor the power system itself to look for changes in current consumption.

Disconnection. Accidental disconnection is another very frequent cause of problems. Not all connectors lock into place, and even when they do, the locks or screws are not always fastened or tightened. This is an acute problem in small offices that do not use a network to share printers or modems between PCs. The correct solution lies in the use of an appropriate network, in which the discipline of screwing all connectors home can be enforced. There are now networks suitable for the very smallest of installations, and they are priced accordingly. The disconnection problem then transfers to network cabling, which can be more effectively managed.

Tampering. There is the possibility that an intruder might tamper with a piece of equipment in order to modify it for his or her own ends or simply to bypass security. This is on the whole a very unlikely form of threat, but the possibility is greater if the way to bypass the security is obvious. Again, out of sight is out of mind: it is best if there is nothing observable to provide temptation.

Monitoring. There is the possibility that someone might monitor (view) the data, either by simply looking at a terminal or other device designed to display data, or by attaching wires to some point at which the signal can be intercepted. Depending on the type of signal at that point, it may be very easy or very difficult to detect this kind of unauthorized monitoring. The data in telephone lines and the very common RS232 serial links between computers are among the easiest to monitor unobserved. The key here is again not to expose potential interception points to view.

5.1.1 Points of Entry

The points at which the signal enters or exits a piece of equipment are the most vulnerable. It is important to use the correct connector in every case; shielded connectors with lock screws are available for most types of equipment and should always be used unless there is a good reason not to. It is often possible to arrange for connectors to be located within the equipment case so that a tool is required to undo them. Most communications equipment should be housed in racks with locking doors.

The point from which a signal leaves a building is also a relatively weak link, and its selection should not be left to the discretion of the installation technician.

5.1.2 Buildings

The building's design and the area immediately surrounding the building can greatly affect the security of the installation. Can all windows be secured? Are there dark areas outside that give an intruder cover? Is there an effective alarm system? Part of the building is often restricted to personnel who have a security clearance; it is worth remembering that this clearance should only be given to those whose trustworthiness has been checked, not to all those whose job requires access.

5.1.3 Telecommunications Areas

Within the building, an area should be selected to house the telecommunications equipment. The criteria used in the choice of area should be a combination of convenience (cabling must be run from this area to most of the building) and security: Can it be made secure? Access to this area should generally be restricted. The telecommunications equipment should be mounted in lockable racks or cupboards. This placement should also apply, as far as possible, to individual modems, although in this case the cupboards may be mounted close to the PCs or terminals they serve.

5.1.4 Terminals

Terminals attached to a large multiuser computer (nowadays less likely to be a mainframe and more often a "midrange" or "supermini" system) normally use serial asynchronous or synchronous (3270) connection protocols, which are particularly easy to tap into and monitor; but secure terminals (where the data passing between the computer and the terminal are encrypted) are very rare. It may therefore be important to ensure that the cables from computer to terminal remain within the restricted area or are otherwise protected.

PCs used on a network are generally similar to terminals, although good network management is necessary to ensure that users do not squirrel data away on a local disk, thus undoing all the protection the network can offer. Laptop and notebook PCs are often used in this way deliberately, and they are particularly prone to theft. PCs in this class should always be protected by hardware or software security devices.

5.1.5 Modems

Modems operating at low speeds (up to V.22bis or 2,400 bps) are also rather easy to monitor. The connections to most modems are standardized, and the protocols usually follow a very widely used standard (the Hayes "AT" standard). One of a small number of common software packages is frequently used for control of the modem, and the number of configuration parameters is small enough that they can usually be arrived at by guesswork or trial and error.

Modems used for dialing out can be hijacked by unauthorized applications, but this is unlikely to represent a serious threat if the modem itself is not secured. Any modem that accesses the public network must by law use a standardized, and completely insecure, connector; a telephone line can be monitored simply by wiring another (preferably high-impedance) device alongside it: another receive-only modem is then sufficient to monitor the data passing across the link.

In PCs, internal modems are slightly more secure than external ones (the signal cannot be intercepted between the PC and the modem), and leased lines are more secure than the public network. But the only way to make the modem itself secure is to have an encryption and mutual authentication scheme between two modems that "trust" one another. Such schemes are usually implemented using a smart card or other device that can be removed when the modem is not in use.

5.1.6 Multiplexers and Network Devices

Within a building, terminals and incoming modems are often connected to a multiplexer, which is a more efficient way of making wired connections to the central computer.

With a LAN, there are other devices, called *bridges* and *routers*, that do the work of concentrating signals into a central point. These and other network hardware are dealt with in more detail in Section 5.3.

Multiplexers, bridges, and routers are more critical to the reliability of the network than other hardware, since a failure can make all or a large section of the network inoperable. Redundancy should always be considered in network design, but is often not taken into account for multiplexers.

Modern high-speed "intelligent" devices are more difficult to tamper with or monitor unobserved, and are at least as reliable as older, slower designs. Devices such as statistical multiplexers, which depend on communications between the two ends, should be used in preference to simpler "time-division" multiplexers.

5.1.7 Encryption Devices

Encryption devices actually fall into two categories: those that really do encrypt data in real time, and those whose main function is authentication. Although in principle an authentication-only device could be disconnected after an initial dialogue, this would not be possible with a well-designed encryption scheme.

The actual devices will again be discussed in more detail in a later chapter, but in relation to the threats discussed above, we can say that these devices are more prone to being stolen or disconnected than to losing power, being tampered with, or monitored. Most encryption devices are designed in such a way that they cannot be copied or altered.

Figure 5.1 shows the main points in a computer network where the hardware is the source of risk.

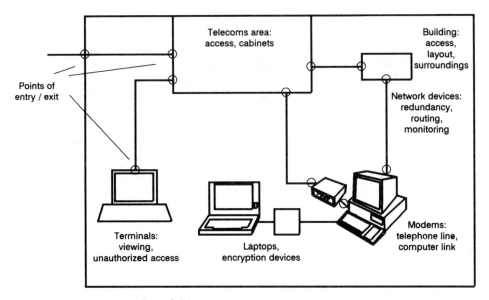

Figure 5.1 Hardware points of risk.

5.2 SOFTWARE

With hardware, we are mostly concerned about a loss of function: the item just stops working, even if only temporarily. With software, the problem can be much more complex, because software can stop working, do something different from that which we intended, become corrupted, and even affect other parts of the system while seeming to do its own job perfectly well.

For normal, tried and tested software, the results are rarely as obvious as with hardware faults, but can be just as serious. One of the biggest problems can be the time required to locate, diagnose, and correct software faults. Software problems can be accidental or deliberate, but they are nearly always caused by people. They can arise at almost any stage in the program's life:

- Specification: Bad or inadequate specification is probably the most common single cause of software errors. It usually means that some condition has been left undefined; the specifier just did not think of this situation.
- Choice of tools: Writing a program involves choosing a methodology, a language, and possibly other software tools, such as a library of related functions. It is essential that they are correctly chosen for the task; languages and methodologies such as MALPAS, OBJ, VDM, and Z are often thought to be the preserve of a few specialists, but they may be necessary for secure applications.
- Design: The software design and coding stage clearly offers the greatest opportunity for the introduction of errors or even of deliberate paths that may subsequently be used for nefarious purposes. While the proper choice of methodology can reduce these risks (there are some very secure techniques, such as voting systems, which will be discussed in Chapter 6), they can never be eliminated. Any significant software always contains errors.
- Testing: Because the design phase always introduces errors, it must be followed by a testing stage in which all functions are tested as comprehensively as possible. For data communications software, the design of software tests is a very complex process in its own right. It is likely to involve the creation of a testing environment or simulator.
- Run-time environment: Because software can be affected by other software, by the hardware on which it is running, and by any other systems it interfaces with, even apparently correct software can break down or misbehave at run-time. It is then debatable whether the fault was caused by the hardware or software. In general, most people feel that if the software has not itself been corrupted, it should correctly identify the problem, even if it cannot recover from it.
- Application: The way a user applies a piece of software affects its security. Systems programmers and other privileged users in particular have the opportunity to misuse a software function for their own ends.

The problems most likely to be encountered vary according to how a piece of software was acquired (was it a standard package, tailor-made by another company, or was it written in-house?) and with the type of software. Different conditions apply to operating systems, standard packages, and application software.

5.2.1 Operating Systems

The trend away from mainframe and terminal networks and toward smaller, distributed systems has resulted in much more user-friendly, flexible systems, but this trend is not always so good from a security point of view. Whereas the risk from a security breach is greater with the larger system, the protection afforded by the operating system is also usually much greater. The following factors have to be taken into account.

User-Friendliness. This is an important differentiating factor among operating systems: a system that is easy to use results in fewer mistakes compared to one that requires the user to remember long command strings or one whose screen is filled with tiny buttons (and where key commands such as "send" and "receive" are always next to one another rather than being separated by some more innocuous buttons). At the other end of the scale, a system that is too instinctive and very easy to use is also likely to be very easy for an intruder or hacker to break into.

Stability. Even though the operating system market is populated almost entirely by very large companies with considerable resources, some of the products they produce are rapidly shown to be "pups," requiring a major "cleanup" release or just replacement. New releases of operating systems are to be avoided where possible: if you are tempted by the features of a new operating system, wait until the market has had time to form an opinion, and read about others' experiences before purchasing it.

Security Services. The range of security services (e.g., file protection, access control, and encryption) offered by operating systems varies widely. Within PC operating systems in particular, the level of protection offered is minimal, whereas mainframe system services are usually very dependable.

Interprocess Protection. A similar situation exists with interprocess and interuser protection. With most large systems (from superminis upwards), it is taken for granted that one process should not be able to affect another except through the operating system itself, whereas smaller systems offer minimal protection. Within the mainstream PC market, the DOS family of operating systems offers little or no protection for users or processes. The Novell Netware family is significantly better, particularly for the small minority of users who make use of all its security features. But the only current system that should be regarded as acceptable for a security-minded installation is Windows NT, which has been designed from the outset as a secure operating system.

5.2.2 Standard Packages

Standard packages cover a wide range from very stable and dependable applications, such as the popular spreadsheets, word processors, and accounts packages, to highly specialized applications designed for niche markets by one-man systems houses. But even the stable applications can sometimes interact with or make assumptions about some other software, hardware, or operating environments that result in undesirable consequences.

The first thing to do with a software package bought off the shelf is to ensure that the package was designed for the environment in which it is to be run. Often the level of detail given in the sales literature is inadequate for this purpose, and more detailed specifications are required. The detailed specifications can usually be found in the user's manual *after* you have bought the package!

Is it suitable for the application? This may be a question not only of minimum specifications, but also of the type of company it is designed for, national standards, and possibly even business norms that vary from industry to industry, even within a country. Does the supplier offer an appropriate level of support?

Lastly, we come to the subject that system managers dread: viruses. How can we be sure that the software does not contain a harmful virus? Sealed packages bought from reputable sources should be reliable, but cheap software for PCs, Apples, and similar systems must always be regarded as suspect. Virus checkers are not completely reliable, but it is certainly a sound precaution to pass every disk entering the system through a trusted and up-to-date checker.

5.2.3 Communications Software

Communications packages are both a boon and a nightmare. While benefiting from the development and, one hopes, experience that has gone into any purchased package, the user must also recognize that every package has been developed in and probably for a specific environment.

The range of conditions the package can handle and the errors it can recover from varies according to this environment. For example, some packages are excellent in a bulletin board environment, using a wide variety of direct-dial modems, retrieving files, and recovering from noisy lines and relatively long delays. Others have been developed for use in corporate networks; these are much better at handling interfaces between analog leased or dial-up lines, synchronous protocols, and packet switching systems, for example.

Technical staff often prefer a low-level package that gives full control of all parameters (such as the ubiquitous Columbia University Kermit), whereas end users probably want a package that only performs the specific task they want.

In a secure installation, low-level communications software should be discouraged wherever possible. Particularly dangerous is the ability to download and upload complete files.

5.2.4 Application Software

As we mentioned earlier in this chapter, the main danger with application software is inadequate specifications. This leads to unknown conditions and hence to unpredictable behavior.

The result of this may be a complete loss of function; it may recover itself or require manual restoration. Data may be destroyed or altered, and even other programs may be affected. Thorough testing is the only solution, and even this will never eliminate every possible error.

Properly written application software will prevent misuse at run-time or render it difficult; but again this requires comprehensive specifications, not only of the software itself but also of the management controls surrounding it.

5.2.5 Unauthorized Software

One of the biggest problems facing many system managers is the growth of unauthorized software being used on a network. The existence of pirated copies of software on a network can lead to support problems and even prosecution, but is unfortunately tolerated even by some large concerns. Organizations should be prepared to run regular audits of software in use.

In a data communications environment, users may be able to download any of the thousands of programs freely available through bulletin boards. While this is great news for the computer hobbyist, it is very dangerous indeed on a corporate network.

Some of this software may consume excessive resources, while software downloaded through bulletin boards and electronic mail systems is the main source of viruses. The best solution, if practical, is to avoid having any software capable of downloading files (or receiving downloaded files) and making sure that there are no floppy disk drives accessible to users for loading software. Failing this, a comprehensive virus checking system is required on the network. As we will see in Chapter 10, there exist products that can guard a whole network against viruses introduced in this way.

5.2.6 Open Systems Interconnection

Most software systems today pay at least lip service to the OSI principle, which is supposed to ensure that devices and software from different suppliers can work together. In practice, very few systems are fully OSI compliant, but the standard is still useful for the structure of layers it has introduced (shown in Figure 5.2), which does make the range of compatible systems wider.

For an independent system, there are likely to be security services at three layers in the OSI model:

- At the physical layer, where hardware encryption devices are used;
- At the transport layer, where software encryption, access control, and generalized authentication services may be provided by the operating system;
- At the application layer, which should meet the specific security requirements of the application. In general, the higher levels are able to provide much more targeted security, which is likely to be more appropriate and cause less overhead.

There is also a separate specification for a Security Architecture within the OSI standard, and this will be referred to in Chapter 10.

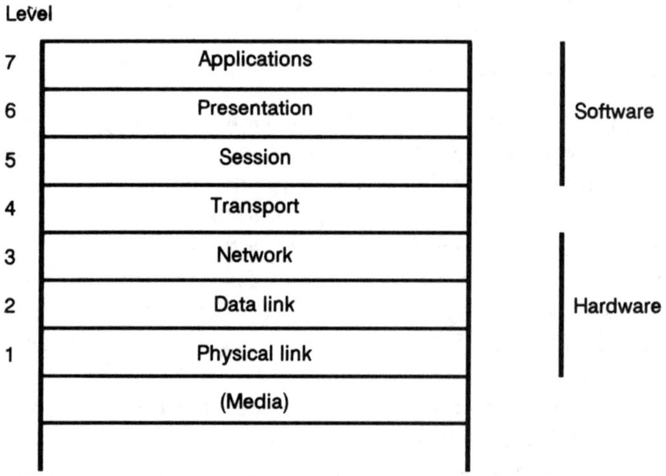

Level

Level		
7	Applications	
6	Presentation	Software
5	Session	
4	Transport	
3	Network	
2	Data link	Hardware
1	Physical link	
	(Media)	

Figure 5.2   Open systems interconnection reference model.

5.2.7 Databases

Many different types of database will be considered in Chapters 13 to 16, which deal with applications. But the existence of a database on a data communications system is a risk. The most likely risk is that some unauthorized person might view or copy the data. In principle, this should be prevented by access controls, as described above, so that no one can see data to which he or she should not have access. Where the data are regarded as particularly sensitive, however, some extra precautions are often desirable, such as encrypting the data on the file, so that they can only be read with a legitimate application, logging all accesses to the data, or using removable disks and removing them at night.

The key management control, however, is to have a database administrator whose job it is both to keep the data secure and to ensure that it is correctly backed up and available for recovery. This person need not be a computer specialist, but should work with the system supervisor to ensure that network access levels and privileges are correctly set.

5.2.8 Redundancy, Recovery, and Backup

A critical area for the security of any computer system is the extent to which it can recover from faults and problems. Although this is more of a procedural than a software aspect, the provision of the following items must be specified and designed into the software:

- Redundant paths (alternate ways of achieving the same goal);
- Recovery procedures (to isolate the fault and restore normal operations);
- Backup procedures (so that only the minimum amount of data is lost in the event of a problem).

The damage caused by any incident will be minimized if these items have all been thought out and written down in advance (and followed: most installations have a backup procedure but not all follow it religiously). The action of working these aspects out on paper beforehand can often alert staff to weaknesses in the system, and it is always better to have thought them out without the stress of a major incident in progress.

In a data communications system, the provision of redundancy is normally a part of the network design. Recovery from common faults can often be automated, but there is often a manual fallback action, such as switching to a reserve modem. As many generations of backups as is realistic should be kept: the optimum combination of archive files, complete and incremental backups, will depend on the size and nature of the system, but a good backup system will impose very low overhead.

Figure 5.3 lists the main points of risk and controls applicable to the software elements of the system.

5.3 NETWORKS

Networks use a combination of hardware and software, and most of their weaknesses are similar to those mentioned under the hardware and software headings.

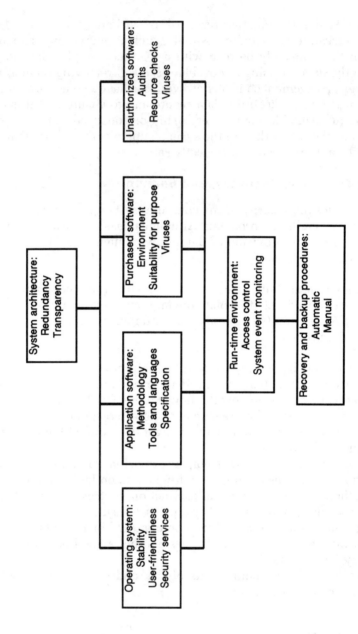

Figure 5.3 Software points of risk.

Cables. Network cabling needs to be a compromise between the needs of redundancy, ease of maintenance and reconfiguration, and monitoring. Most standard office networks use either the Token Ring or Ethernet standards, although larger systems may use the fiber-optic-based fiber distributed data interface (FDDI) for links between floors or other parts of the network. Under the original "thick" Ethernet standard, the network had to be disconnected in order to add or remove a device, which was inconvenient but good for security. With Token Ring and the newer Ethernet standards, disconnection is not necessary. The current choice for structured cabling of networks is to use plain twisted-pair cables following the 10BaseT Ethernet standard. The more intelligent controllers, known as *hubs*, used by this system are capable of detecting most cabling faults and attempts at tampering with the network. With the other systems, it is advisable to install a network monitor to provide these facilities. The choice of cable routing should also take into account the needs of maintenance and security.

Connectors. Some connectors allow the network to continue running while connections are made or uncoupled; others do not. The choice between these types of connectors will again depend on the balance between convenience and security, although with good network monitoring equipment and software in place, the former will always be preferred.

Bridges and routers. Like multiplexers in an asynchronous terminal environment, bridges and routers are more vulnerable just because more traffic passes through a single point. They would therefore be a target for anyone wanting to disrupt the network or perhaps attempt to connect into it or monitor traffic on it. *Bridges* are relatively fixed in their function, but *routers* offer more flexibility in translating protocols and network standards. *Gateways* have the greatest potential for interconnection to and from external systems and need to be treated with the same respect as the computer system to which they are connected.

Other people's terminals. Any terminal on the network is a potential source of risk. As we mentioned in Section 5.1.4, the characteristics of the terminals themselves can be chosen to minimize the risk, but operators of WANs must be aware that other terminals can often be connected to the network; PCs used as terminals are powerful tools for the would-be hacker. Provided that all the other controls mentioned above are in place, such terminals should not be able to cause any other harm.

Other computers on network. The same is not true of other computers connected to the same network. Under the OSI model, the connection between the computers is at the physical layer, so that someone with access to low-level functions on the remote computer is also likely to be able to gain this level of access to your computer. Although software controls may still protect your system to some degree, real protection for your system is likely to depend on hardware encryption on the link, controlled by application-level software—a breach of the OSI principle but nonetheless necessary.

Figure 5.4 shows the key points of risk in a network environment and the associated threats.

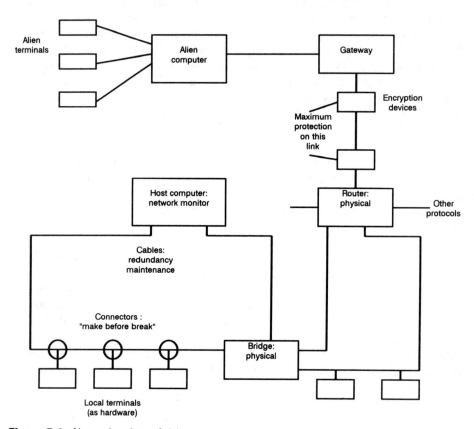

Figure 5.4   Network points of risk.

5.4 COMMUNICATIONS SERVICES

Nearly all data communications systems make use of some external communications services: the public switched telephone network (PSTN) or an X.25-based packet-switching network are the most common today, although integrated services digital networks (ISDN) are also becoming affordable for many applications.

Buying into a service always involves certain commercial risks:

- The risk of the trading partner going bankrupt, withdrawing from that line of business, or selling the business to an unwelcome partner;
- Possible conflicts of interest: the risk of the service provider doing business with a competitor of yours;
- Defining and enforcing levels of service: a particular problem when the service provider is a national monopoly.

A data communications user faces the following specific risks.

Local loop. Today, the link from the local exchange to a customer site is still most likely to be a copper analog link passing through a wide variety of junction boxes, many of which are accessible to the public and are frequently disconnected by mistake by engineers dealing with another customer's problem. These copper local loops are extremely easy to monitor, and they should be avoided in favor of fiber-optic links wherever possible.

Exchange. Most local exchanges are physically reasonably secure, and in European and North American countries, the switching equipment is now almost completely digital, which has increased reliability dramatically. Nevertheless, exchange problems do occur and usually involve some minutes or hours of downtime. It is worth knowing in advance what options your local exchange supports: it may be possible to have alternative routes set up to minimize the effects of this type of problem.

Interexchange circuits. These circuits now are also almost exclusively digital, using fiber optics (and very occasionally satellite links). There is a high level of redundancy with these links, except in those countries regarded as seriously underprovided with telephone circuits. Downtime, when it does occur, is usually of very short duration. Users of international circuits in particular need to be aware of the grade of link they are likely to be using: some calls may be switched to much lower grade circuits than those of calls close to home. The main effect of this is likely to be that modems will not work at their highest speeds.

Switching centers. These are the telephone company's hubs, where interexchange lines are switched. Although they are very heavily protected, a failure, fire, or attack here could bring down the whole network, including all redundant links. If this would be disastrous for the business, it may be worth considering whether another path, not using telephone lines at all (e.g., satellite technology), should be installed as a standby.

VANs. Many data communications networks make use of a VAN. It may simply be a more intelligent switching center, perhaps offering local call access from anywhere in the country, or it may provide extensive processing facilities. The store-and-forward nodes used in EDI and electronic mail systems are examples of this type of VAN. Using a VAN reduces certain types of risk while introducing others. Most VANs have built-in redundancy, not only for the telephone lines and other communications circuits, but also for the central processors and data storage. They have 24-hour operators and good backup and recovery procedures (you still need to verify this for smaller operators!). On the other hand, the number of people who can gain access to the system as a whole is greatly increased; even with good software controls, the risk of a breach is higher than for a stand-alone system. A hacker may be able to remain anonymous to the system while attempting to break in to it. The only answer to this is to ensure that all communications with the VAN are handled by a secure process within the host computer, and that no files or other data are transferred to the main storage until their authenticity has been checked.

Not all of the points discussed above will be relevant or at risk in any given situation. But it is worth being aware of the different sources of risk and the management controls that can be imposed to minimize them before looking in detail at the tools that may be used.

Application-Specific Risks $\qquad$ 6

The levels of the various types of risk, and the potential damage caused by a breach, vary greatly according to the application. Some risks are very specific to certain types of application.

6.1 REAL-TIME CONTROL SYSTEMS

Many computer and data communications systems are put to use controlling equipment in *real time*: the equipment acts upon the computer's output immediately, without any delay. Almost every piece of complex equipment today is controlled by a computer in this way. A failure of the computer controlling a steel mill or oil refinery is serious enough; when the computer is controlling a train, aircraft, or space shuttle, the results may be even more severe.

In this type of situation, a single computer or data link is not allowed to act alone. Usually the computer instructs some more simple electronics (which are assumed to be more reliable). The simpler hardware rejects or limits the computer's output if it is unreasonable, and guides the equipment back to a safe state if the computer fails altogether. This concept of *fail to safety* is critical to all real-time computing. When this control is too coarse, multiple computers may be brought into play, each of which calculates a result, and there may be a *voting* system to determine which output is accepted. The multiple systems may each use a different logic in order to avoid *common-mode* errors, where all the systems have the same error.

We mention these systems first, although most readers of this book will not be directly concerned with computer control as such, because the techniques of limit checks, voting systems, and fail to safety should probably be applied more often to commercial and other systems. They are the equivalents of management controls and are implemented automatically.

6.2   BANKING AND FINANCIAL TRANSACTIONS

6.2.1   Principles

Data communications and computer security are inextricably bound up with banking systems and financial transactions. This is where the effect of a security breach can most obviously be translated into a financial loss for the company and a gain for the person breaching the security. Even human mistakes in this area can have extremely serious consequences for customer confidence and may be difficult to recover from. Banking laws vary considerably from country to country, but there are some common principles:

- One of the most important principles is confidentiality. The details of a financial transaction should not be passed to any third party, and even within a bank a need-to-know principle should apply.
- It is essential to authenticate the originator of a transaction; traditionally this was done through a signature or multiple signatures, and this is often the only proof that a court will accept. However, banks and other institutions will often accept instructions given in person. The City of London was for many years held together by the principle of "my word is my bond." In an electronic age, other methods have to be found to warrant the authenticity of the originator of a transaction. Normal passwords can be "borrowed," and a single electronic "key" such as a smart card can be stolen. If an impostor can successfully impersonate the genuine signatory, then he or she can initiate bogus transactions for his or her own profit. One well-publicized breach of security affecting the British CHAPS interbank electronic clearing system bypassed the electronic controls simply by using forged letterhead and signatures.
- A paper document is often considered superior to an electronic document in the eyes of the law, as is a written signature to an electronic one. This raises the specter of a perfectly valid electronic security system being overruled by an insecure paper one. It is therefore important that the electronic system actually prevents the breach from occurring, or traps it within a very short period, so that it can be corrected without recourse to the law. Speed is of the essence in any monitoring scheme.
- It must be possible to trace every transaction for auditing purposes. In an international and electronic environment, this may mean relying on a wide range of sources; there is insufficient coordination between systems, auditors, and regulatory authorities to enable the creation of a leakproof system. People contemplating fraud often become aware of these loopholes and see them as opportunities; they are usually closed after the event.
- Banks and other financial institutions rely heavily on calculations of exposure (the amount of money they have outstanding in any given position

that is not covered or offset by some other position) in order to limit their risks; their internal systems are often biased toward measuring these exposures. In many situations, individuals have personal limits, which may dictate the maximum size of transaction they can undertake or the total size of a position in a trading environment. Some people have found ways around these limits, while others have been able to change them in order to carry out their fraudulent activity.

Bank staff are generally placed in an exceptional position of trust. While it is possible to use both procedures and electronics to perform constant checks on them, such checks run the risk of being counterproductive if they make staff feel that they are not trusted. As we mentioned in Chapter 3, an atmosphere of mistrust is itself a security risk.

6.2.2 Nonbank Transactions

A large and increasing number of financial transactions take place outside the banking system. Companies manage their own finances by transferring money between subsidiaries or make credit arrangements with other companies without using the bank as an intermediary. These transactions are not subject to any specific rules and may not be subject to any controls until an annual audit many months later. These transactions must be recognized for what they are and subjected to controls as though they were external transactions.

6.2.3 Credit Cards

Credit cards also operate to some extent outside the banking system, although only banks can normally issue them and the networks (Visa, MasterCard/Europay) that run them are in turn owned by banks. They face particular problems in that an individual transaction is often very small, so the security system cannot be allowed to impose too heavy an overhead. The magnetic stripe cards used in most countries are relatively easily copied, at least as far as the electronic part is concerned. Supplementary checks must therefore be performed in order to prove the identity of the cardholder.

The most common checks are signatures and personal identification numbers (PIN). Both have drawbacks: signatures are rarely adequately checked, signature strips on cards can be replaced, and customers often have difficulty remembering PINs and therefore write them down. PIN checking also normally requires an on-line link to a host computer or a smart card. These techniques will be discussed in more detail in Chapter 14.

Actual practice varies widely from country to country. In France, all bank cards are smart cards, and PINs are used for all domestic transactions using domestic cards, normally off-line when below a floor limit. In the United Kingdom, signatures are invariably used in retail outlets, either off-line or on-line, but PINs are used at automated teller machines (ATM). In the United States, credit cards are operated with signatures, but debit cards use PINs. Authorizations are almost always on-line unless the retailer is prepared to take the payment risk.

6.2.4   Financial EDI

Trade payments are increasingly made by EDI, which also offers some scope for fraud and error, although in both respects probably less than with the manual systems they replaced. What EDI does do, however, is increase enormously the scope for checking the payment half of the transaction by a form of double-entry accounting. We will consider this aspect again in Chapters 13 and 14.

6.3   ACCOUNTING SYSTEMS

A company's internal accounting systems reveal a lot about the company. In addition to the identity of trading partners and the volume of company business, they give very detailed information about the financial status of the company.

Errors or deliberately falsified accounting information can be very harmful to the company; it may spend money it does not have, make payments it does not owe, or simply give wrong information to shareholders or regulatory authorities.

From a data communications point of view, it is unusual for an accounting system to be accessed remotely. So provided that the operating system has a good protection system, physical controls should be adequate to prevent unauthorized access. In smaller companies, however, or with less secure operating systems, there is a significant risk of unauthorized personnel at least being able to view accounting information. This information may be regarded as useful to a competitor and is a frequent target for casual industrial espionage.

The biggest risks in this area are those relating to payments: in particular, the ability to set up new accounts and authorize payments. The danger to companies of their financial information being leaked is often exaggerated, although it may prevent a company in a difficult financial situation from recovering.

Audit trails must be comprehensive and yet easy for internal and external auditors to use. Too often, computer audit trails remain unused except in extreme situations. The audit trail must itself be secure: difficult to alter and unlikely to be destroyed or corrupted. It is, however, worth adding that too much security can also be a problem in an accounting system. The security should not have the effect of making the system less transparent or the results less clear, allowing loopholes and anomalous results to go undetected for a long time.

6.4 CUSTOMER DATABASES

Most well-run companies now rely heavily on their databases of customers and prospective customers. These databases are used as links between the different parts of the computer system, as a communications tool between staff, and simply as a filing system that can be easily accessed. Corruption or copying of these databases is a serious risk for most companies. If the customer database becomes unavailable for any reason, large parts of the computer system may be unable to work. If the database is corrupted or altered, and the problem is not noticed in time, a whole range of customer service problems may ensue.

Customer databases are particularly useful to competitors, and it is very common for employees leaving the company to take a copy of the database, often the day before handing in their notice. Even where some or all of the data are in coded form, the employee is likely to have enough samples of the true text (on business cards, in notebooks, or in memory) to be able to work out the coding system.

It follows that for customer databases, a rigorous backup system is particularly important. Access controls to this system should be just as stringent as those to the financial system, and where possible data should not be held in a single file, but distributed around the system so that each application has access only to the files with the data it needs to know.

6.5 LEGAL AND CONTRACT DATA

Lawyers and solicitors are traditionally very wary of computers in any form. The courts in most countries still only accept computer-generated documents or data as a lower grade of evidence. This cautious attitude can lead to a lack of knowledge of computer processes, which is dangerous in the context of their own business. Legal documents, and the data underlying them, can now be transmitted between offices and between parties and marked up on screen, and new drafts can be generated within hours. Some large offices have very extensive and powerful networks and use on-line links or CD-ROM (compact disk read-only memory) databases to check legal sources. Where traditionally lawyers and solicitors have worked in single, often large offices, they are now frequently part of a larger network encompassing many offices.

Confidentiality and accuracy are essential in the lawyer's trade. Much of the lawyer's work is sensitive, and in some cases such as corporate takeovers and mergers, disclosure of even small details could result in serious financial claims.

Chief financial officers, commercial departments, and others involved in contract negotiations face many of the same problems. They often have to use a word processing system shared with a large number of nonsensitive applications. Creating dividing lines between these systems is often difficult.

There are many areas where technology could be used to help ensure contract compliance (e.g., in the regular exchange of agreed-upon data). This is rarely specified in contracts, and still less often in a form that would be regarded as secure. It would be difficult for any standard form of risk analysis to take into account the risks arising from a lack of data communication, but any strategic review should consider these actual risks and consider whether technology could be used to reduce them.

6.6   INTELLECTUAL PROPERTY

For some companies, design is a large part of the value of the company. Design time frames are becoming ever shorter, but greater resources are continuously required to meet them. This means that a competitor who can obtain access to all or part of a new design can derive a large advantage from it. Often the mere existence of a design must be kept secret.

In many environments, the full details of a design can only be stored on a computer-aided design (CAD) system. Where designs have to be passed from one company to another (e.g., to a subcontractor), there may be problems of compatibility between the systems, so that an interchange format has to be agreed on. This interchange format is likely to be a published standard, and of course even the proprietary standards used by different CAD packages are readily available.

Ad hoc transmission of design data between companies is therefore a particularly weak area for security and should be avoided whenever possible. There exist now several EDI systems suitable for technical data, and they have usually been designed with adequate security features. These systems include the STEP (Standard for Exchange of Product Model Data) system for mechanical and product design data, and the EDIF (Electronic Design Interchange Format) for electronic design data. STEP and EDIF files are normally exchanged using a commercial VAN such as the IBM Information Network or AT&T EasyLink. Messages usually include a check sum, which is used, together with the time, date, and message number, to authenticate the message and prove that it has not been altered. These techniques are explained in more detail in later chapters.

Research data, such as the results of a laboratory test, can be equally sensitive, and they do not come in any standardized format. They can, however, usually be coded in such a way that the results on their own are meaningless.

Designs that have not been patented or published are vulnerable to being copied and possibly patented or published by someone else. Where patent protection does not apply, as is the case for most software, users rely on copyright, and it is important to be able to prove the date when a document or file originated. We will see later that electronic signatures can, and usually should, include a date and time field. In some cases, it may be important to include the date and time when the file or document was created, as well as its last update or the time of transmission.

6.7 SUBSCRIPTION SERVICES

Subscription services include not only the large and growing number of on-line databases, but also the networks that give access to them—any situation in which a large number of relatively uncontrolled users has some form of access to the system. In principle, anyone who connects to such a network, whether as a subscriber, information provider, or network provider, is at risk. Hacking, or unauthorized access to a computer system, is the most publicized risk in this case. The hacker may have legitimate access to one part of a system, but exploit a protection weakness to venture into other areas, or the intrusion may not be recognized by the system at all. Hackers may intrude harmlessly and leave little or no trace. They may simply read files, or they may leave messages to show that the security has been breached. The greatest damage is of course done when files are deleted or altered, or when the hacker leaves a virus or other monitoring program.

There have only been one or two cases in which hackers have been able to use the system for more pernicious ends, such as committing fraud. In these cases, the hacker has nearly always had some prior access to the system, probably as a programmer or computer operator, and has left a *hook* that he or she was able to exploit.

Legitimate users have been able to do considerable damage to such networks; for example, there has been more than one case of electronic mail networks being jammed by users sending long files or programs to everyone on the network. Such programs and systems must be extremely careful to control the resources available to each user and cannot rely on operating system protection.

Unless you are extremely confident of the protection afforded by your operating system, you should avoid allowing outsiders to dial in to your computer system and log on using a password only. Almost every password system has been broken or circumvented, usually because the hacker has obtained a legitimate password. If the password has a sufficiently high privilege level, the hacker may then be free to initiate many other processes and hide his or her traces. Where dial-in access is required (e.g., for field sales staff), consideration should be given to using a dial-back technique.

Those who actually provide information or network services cannot avoid giving outsiders the right to connect to their system. They will obviously want to make use of an appropriate access control system, but should also normally give access only to a copy of the data, extracted from the main file at intervals and lodged on a separate system.

Network service providers, although continuously exposed to security risks, are not usually themselves the target. They may, however, suffer customer relations and service problems if their customers experience a security breach. They have the greatest chance to control the security on the network, and can impose conditions on the hardware, software, and procedures to be used by their customers in order to maintain security. Their own operations and data (including customer files) are, of course, subject to the same problems and risks as those of any other firm.

Special mention needs to be made of the Internet, a supernetwork that connects together a vast number of other networks and individual users. It originated in the research and academic environment in the United States, but has now extended to cover almost every type of organization in all parts of the world. The Internet is widely used for its bulletin board and electronic mail services, but many other forms of service are also offered, including business "for sale" and "items wanted" advertisements, home shopping, information services, and interactive forums. The confidentiality and origin of mail and other messaging services on the Internet cannot be assumed unless specific precautions have been taken.

The Internet is a very free and open network; no one controls it and there is no tariff for using it (although most individual users access it via a subscription service). Nevertheless, it is used today for serious commercial traffic, and since mid-1994 it has been possible to offer commercial services on it directly. These services have to look after their own security, since the network does not have any security infrastructure. Users connecting to the Internet must also ensure that they are not putting themselves at risk from hackers or unauthorized users. Internet usage is by nature somewhat ad hoc, and it is difficult to program specific checks into application software. It may be necessary to limit the scope of any possible damage by using a nonnetworked PC for Internet access.

The Internet has also been the battleground for several disputes over the rights of cryptologists to use, and in particular to export, encryption software and keys. We mention this further in Section 6.10.

6.8 RADIO AND SATELLITE NETWORKS

For a variety of reasons, many networks use radio rather than fixed connections. The users may be mobile (delivery vehicle drivers, salesmen, or warehousemen), the equipment may be moved often (in some offices), or the environment may make fixed cabling impractical or uneconomical (retail shops with concrete floors or links between offices across a street). For longer distances, or to reach very remote sites, satellites are increasingly used.

All networks suffer from some errors: on analog telephone lines, the error rate can be as high as 1 in 10^3, although 1 in 10^6 is more common. Leased and digital lines are much better. Radio networks can match these figures under good conditions, but under bad conditions they can suffer from a wide variety of problems, such as fading (temporary loss of signal), multipaths (two signals reaching the aerial at different times), crosstalk (interference from other radio channels) or jamming (strong interference that masks the original signal completely—accidentally or deliberately). There are techniques that handle nearly all of these conditions (except effective jamming), but the problem is to ensure that the conditions are known in advance so that the right techniques can be applied.

Another potential problem with radio networks is monitoring. The signal covers a wide area, and another receiver can pick it up, usually without any noticeable effect. In some situations, this problem can be minimized by using highly directional antennas, but the only completely effective system is to use a *spread spectrum* technique, as described in Chapter 8.

Spread spectrum techniques "hide" any one signal within a much wider bandwidth signal. Someone picking up the signal hears only random white noise and may be unaware that there are any useful data on the channel. Code words and correlation techniques are required to extract the data. The characteristic of making it difficult to tell that there are useful data on a channel makes spread spectrum techniques valuable even when there are highly secure algorithms available and cheaper to implement on the digital side. The GSM (Global System for Mobile Communication) standard for digital cellular radio, on the other hand, uses digital encryption only, since it is only the signal content, not its existence, that is being hidden.

Satellite transmission can suffer from all of these problems, although the environment can usually be worked out in advance for fixed sites at least, and appropriate measures taken. The dish area may sometimes have to be increased in order to give a strong enough signal for good data reception under all conditions. Rain, for example, causes a major drop in signal strength. Satellite signals can suffer interference from neighboring satellites; but with good design and adequate dish sizes, this should not be a problem, although the standard geostationary orbits are now becoming so full that intersatellite interference will become an issue.

6.9 PERSONAL DATA

Privacy legislation in many countries means that special care has to be taken with personal data, above and beyond any commercial implications these data may have for the company, if it is to avoid prosecution and bad publicity. It is wise to have a specific flowchart drawn up for any personal data held within the system or transmitted through it. This flowchart can be checked by data protection officers.

6.10   MILITARY AND NATIONAL SECURITY

"National security interests" have always been invoked by governments, some-times legitimately and sometimes spuriously, to justify keeping data secret. In a computer network, it is more difficult to achieve absolute secrecy, but also, in fact, easier to implement a graded security scheme. Most large countries have teams of cryptologists and hardware and software specialists working on cryptography. Their results are used not only in military applications, but also to maintain the secrecy of much internal government activity.

Although nowadays the principles of cryptography are well enough known for many competent mathematicians to be able to devise their own schemes, governments in many countries reserve to themselves the right to decrypt anyone's data. This means that the description of any new encryption scheme (and any special key generation algorithms but not the keys themselves) must be lodged with a government agency. The U.S. government particularly objects to the exportation of encryption schemes or software.

For many years, DES (ANSI X3.92), described in Chapter 7, has been by far the most common encryption scheme. The U.S. Defense Department, for whom DES was developed, is keen to have an even stronger hold on its successor: a hardware chip known as *Clipper* is the government's proposed standard and it has been the subject of much opposition from security systems developers and users. Clipper is now likely to be restricted in scope.

Military communications depend very heavily on radio. In addition to the techniques mentioned above, modern military radios use a combination of spread spectrum and encryption, often using special key generation *guns* that generate a repeatable sequence of pseudorandom data for use as keys. Access to such guns must be carefully controlled; in particular their *seed keys*, which start a particular sequence, must be distributed securely by separate methods.

Another very special area is that of very-low-frequency (VLF) radio used for communicating with submarines. Here the data rate has to be kept as low as possible (because of the low frequency), and so the overhead imposed by many encryption schemes would be excessive. What is done instead is to use a combination of coding and data compression (with minimal encryption) to send the smallest possible quantities of data.

Table 6.1 lists the application-specific risks and associated issues mentioned in this chapter. This table can only show a sampling, but it should give sufficient pointers for anyone seeking the risks relevant to their application.

Table 6.1
Application-Specific Risks and Issues

Application	*Risk or Issue*
Real-time control	Loss of control
	Over-limit outputs
	Common-mode errors
Banking and financial transactions	Confidentiality
	Originator identity
	Invalidity of computer "proof"
	Audit loopholes
	Personal limits exceeded or changed
	Atmosphere of mistrust
	Transactions outside banking system
	Credit cards (overheads vs. security)
	Trade EDI
Accounting systems	Errors
	Falsified information
	Bogus trading accounts
	Spurious payments
	Information useful to competitors?
	Audit trails usable and secure
	Not so secure that transparency is sacrificed
Customer databases	Corruption
	Effects on other systems
	Copying (especially by employees leaving)
	Backups
	Access rights control
	Distribute data?
Legal and contract data	Confidentiality
	Accuracy
	Inhouse as well as lawyers
	Shared word processors
	Use of data exchange to check contract compliance
Intellectual property	Design details or existence
	Interchange formats: EDI
	Ad hoc data: coding schemes
	Patents and copyrights: date and time protection

Table 6.1
(Continued)

Application	Risk or Issue
Subscription services	Unauthorized access
	Abuse by authorized users: resource limits
	"Planted" hooks
	Avoid or restrict inwards access
	Limit access to a copy of the data
	Network providers: customer confidence
	Internet and similar networks: applications must ensure their own security
Radio and satellite networks	Fading, multipaths, crosstalk, jamming
	Monitoring
	Intersatellite interference
Personal data	Data protection
Military and national security	Right to use and export
	Successor to DES
	Special encryption for military radio
	Submarine communications

Part II: Technology

Encryption Principles

7.1 THEORY AND TERMINOLOGY

Most of the techniques used to enhance the security of a data communications system involve the use of encryption. By this we mean transforming the data so that they cannot be understood except by someone who knows how to decrypt them. Cryptologists call the original data the *plaintext* and the transformed data the *ciphertext*.

The encryption function, or algorithm, usually uses a *key*, which is a further block of data, so that the process does not depend on keeping the algorithm itself secret. The key itself should be changed regularly. The simplest form of using a key is shown in Figure 7.1.

For example, a really simple encryption algorithm would involve adding 1 to the ASCII code of every character: a becomes b, x becomes y, 3 becomes 4, and so on. This is extremely easy to break, and can almost be done by eye. If, instead of adding 1 every time, we add a different number in a fixed sequence: 1, 45, 24, 17, 82, 21, 56, 77, then our ciphertext would probably withstand casual analysis unless it contained long sequences of similar text, such as many blanks or zeros. The number sequence in this case is our key. Of course, practical encryption algorithms are much more complex than this. The scheme described above would be called a *stream cipher*, since each byte can be transformed as it passes through the encryption function. Most practical schemes, however, are *block ciphers*: they take a block of data, typically 64 or 128 bits, and manipulate it before transmission. Short blocks are padded in a defined way.

In most encryption schemes, the same key is used for encryption and decryption (in our simple model, we simply subtract the numbers in the same sequence). Both the sender and the receiver must therefore have a copy of the key. This means that either keys have to be sent from one place to another or there has to be a way of regenerating precisely the same sequence of keys. The second technique is used in *key guns*, which are described in the next chapter, but for

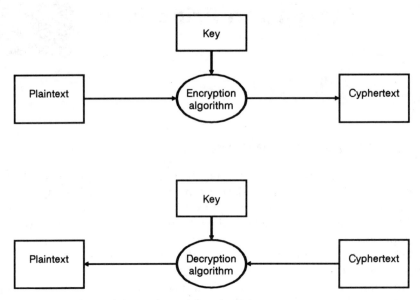

Figure 7.1   Encryption and decryption using a single key.

most purposes keys are sent hidden in the message, probably encrypted by a further level of key: the *key-encrypting key*.

Encrypting one block of data creates another block, normally of exactly the same length. Strong algorithms use a complex sequence of chopping and adding bits, so that even a single bit change in the original plaintext will change many bits in the ciphertext block. Ideally, half the bits will change. The new block can be further encrypted as many times as necessary: provided that each encryption is matched by the equivalent decryption, exactly the same plaintext will emerge at the end of the day.

In a data communication system, the plaintext is reduced to a string of meaningless bits, which can be chopped up into blocks, have control characters added to it, and encrypted and decrypted several times. Provided that there is no loss of information, and that the integrity of the communications path has been maintained, the true data will emerge at the other end.

The number of permutations of the data is directly related to the size of the key. For perfect secrecy, the key must be at least as long as the original message, and it can only be used once. This makes key transmission just as critical as sending the original message, and so it is not really a practical option. What it does show, however, is that long keys are more secure; the trend nowadays is toward very long keys, sometimes as long as 128 bits. It is also important that keys are truly random and unrelated to the previous key. A long key is also more effective when it comes to demonstrating the integrity of the communications systems using MACs, as described below.

If a strong algorithm is used, then the algorithm itself can be made public, and only the keys need to be kept secret. This is the technique most widely used today; in fact, most commercial encryption follows one of two or three published standards, although the techniques used for key management and message authentication vary considerably.

7.2 KEYS AND KEY MANAGEMENT

If the algorithm used in an encryption scheme is known, then it is clearly critical that the key management system is secure. The type of key described above is *symmetrical*; that is, the same key is used for encryption and decryption. It does not matter in this case whether the algorithm itself is reversible, or whether there is a complementary function for decryption. In the case of the popular DES algorithm, it is reversible. Our simple stream cipher above required a complementary function, but the key remained the same.

A symmetrical key must be known at least to the two parties in an exchange, and it is often used by a whole user group. Such keys are sometimes derived by an algorithm using a fixed or infrequently changed key and a common variable, such as the date and time. Special calculators may be used to generate them. More commonly, however, they are exchanged as a part of the setup for a particular transmission or coded message transmission. The key structure will probably consist of a hierarchy, including keys that belong to a whole user group, master keys that are secured in tamper-proof memory in each piece of equipment, keys or PINs for individual staff members or functions, and finally a session key that is encrypted by a master key and then exchanged once all the other checks have been completed.

During the session, all data are encrypted using the session key. At the end of the session or at any critical point, it can be discarded and a fresh key selected. An example of a possible hierarchical key structure is shown in Figure 7.2.

ISO 8732 (which corresponds to ANSI X9.17) describes such a multilayer key management scheme, which is used in interbank communications.

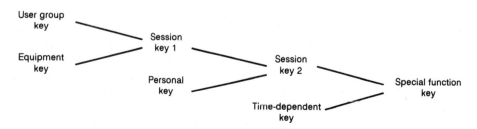

Figure 7.2 Example of a hierarchical key structure.

7.3 PUBLIC KEY SYSTEMS

An alternative approach is offered by the *public key* or *asymmetrical key* systems. In this case, the key used to encrypt the data is different from that used to decrypt it, and it is infeasible to derive the decryption key from the encryption key. (In encryption, it is always acknowledged that there is a remote possibility of a cipher being broken, but it is judged infeasible if the chances of hitting on it accidentally are almost zero and the computer power required for a deliberate attack is greater than anything currently available.) This method depends on finding numbers with specific characteristics to use as keys; unlike the symmetrical key schemes previously described, it is not possible to use a random sequence. In fact, key generation itself requires considerable computation.

There are, however, two key advantages in such a scheme:

- The encryption key can be published (hence *public* key), and the plaintext can still only be recovered by the intended recipient. This makes key management much simpler.
- Because the secret key is secret to only one party, it allows a *digital signature* encrypted with this key to be firmly associated with that party; in other words, it supports a very strong proof of nonrepudiation.

There are also disadvantages: keys cannot readily be changed by a host system. Asymmetrical schemes generally require more processing power for their encryption and decryption, and simple broadcast messages cannot be supported.

For these reasons, public key systems are most often used for demonstrating the authenticity and integrity of a message, after which the main body of the message is likely to be encrypted using a symmetrical scheme. In many cases, where privacy is less of an issue than the commercial and technical integrity of the system, this is not even considered necessary, and the message itself may be sent in cleartext form. This is, for example, how some trade EDI schemes operate.

7.4 MESSAGE AUTHENTICATION

Message authentication in practice has two main purposes: to ensure the *integrity* of the communication and to verify the *identity* of the sender. At a low level, integrity checks typically consist of parity bits added to each byte, which enable single-bit errors to be detected but not corrected. With modern data communications, single-bit errors are unusual; if communications are disrupted, long sequences of data are turned into complete rubbish. Parity bits are still often used, just because seven bits are enough to display all normal printing characters, though eight bits are a more convenient word size for a computer.

Graphical data, programs, and other special files usually need 8-bit bytes, so parity cannot be used.

Check sums, usually the least significant few characters of the sum of the message's characters, are more generally useful. They also have the advantage that they will change with any change in the message content.

Other useful characteristics of a message are the *date and time* it was sent, and its *sequence number* (which helps to identify lost messages as well as corrupted ones).

The check sum, date and time, and sequence number are the most commonly used elements in creating a *digital signature*. Sometimes a PIN or other short sequence is used as well to identify the sender personally. The chosen elements are *hashed*—chopped up and rearranged—and then encrypted. With an asymmetric algorithm, the sender's secret key is used for this encryption; otherwise it will be the common key. For extra security, a *one-way function* is used instead of the hashing process.

When the message is decrypted, the receiving end regenerates the check sum and other elements of the signature and then checks that the signature as sent corresponds with that regenerated. With an asymmetric key, the receiver must first decrypt the signature using the sender's public key. If this produces the right result, then the message can only have been sent by the right party.

This whole process is known as the MAC, and is often referred to as "MACing" a message.

7.5 DATA ENCRYPTION ALGORITHM

The DEA, more often known as DES, is by a large margin the most commonly used encryption standard today. It was published as a standard [1] in 1975.

DES is a symmetrical algorithm that operates on 64-bit blocks of data with a 64-bit key using a repeating sequence of chopping and binary addition. It has a number of very attractive features, particularly the fact that it is completely reversible: decryption and encryption are exactly the same process using the same key. Because of its repetitive nature, it is also relatively easy to implement in hardware, as we will see in the next chapter.

In theory, DES is controlled by the U.S. Government, which restricts exports of encryption hardware and software. In practice, it is so widely used and distributed that versions of it even exist in shareware. Although cryptologists can find weaknesses in it, it is generally reckoned to be adequately secure for most practical commercial purposes today. With existing key lengths, however, it is not sufficiently secure for top secret applications in military and government use; and with increasing computer power becoming widely available, its restrictions will only increase with time. Its most important restriction, though, remains the fact that it is a symmetrical algorithm and depends on secure key management and distribution. This is where the public key algorithms come into their own.

7.6   PUBLIC KEY ALGORITHMS

The best known of the public key schemes is the RSA algorithm, named after the mathematicians who described it in 1978 [2]. This scheme depends on finding pairs of numbers with specific characteristics, which are then manipulated, and a further *pair* of numbers is published as the public key. Decryption actually involves a knowledge of these two numbers and a further number (the secret key), and is relatively complex mathematically—it is nearly always performed in software rather than in hardware.

When the RSA scheme is employed for signatures, it uses the encryption function with the secret key only, and signatures can be tested using the same function, but with the public key.

Although the RSA algorithm is proprietary (and also theoretically subject to U.S. Government restrictions), there are several implementations commercially available, and it is incorporated in many standard products.

Other asymmetrical encryption functions sometimes used for signature purposes include *trapdoor knapsack* and *discrete logarithm* techniques. These are basically one-way (nonreversible) functions and are occasionally used when there is no need to decrypt, but only to authenticate.

Key management with an asymmetrical scheme is much easier, although keys are not normally changed as often as with a symmetrical scheme. Within a user group, public keys may simply be held in a directory. There is specific provision for this in the X.500 directory system used with OSI, and many EDI systems have similar facilities.

7.7   CURRENT DEVELOPMENTS

Encryption is an ancient art. It has been used for centuries for safeguarding the secrecy of government and military communications. It is really only in the last twenty or thirty years, since the emergence of computer communications, that it has become a matter of commercial significance. It is remarkable how much progress has been made since the development of information theory and a key paper by Shannon [3]. Nearly all the principles outlined above have only been accepted during this period.

It is now clear that DES, which has served the industry well, is reaching the end of its life for pure privacy applications, although it is likely to remain in use for many years as a general tool. RSA and other related asymmetrical systems are mostly important because of their value in creating and validating digital signatures, but they are still somewhat cumbersome to use for encryption of data streams.

The time is therefore ripe for the emergence of a new major algorithm or technique. Such an algorithm must be easily implemented in hardware, which favors an iterative scheme along DES lines, but with larger, and possibly variable, key sizes. Key management and key transmission must be an integral part of such a scheme.

At the time of writing, three main contenders have appeared: the Japanese-invented Fast Encryption Algorithm (FEAL) [4], the Australian LOKI [5], and the International Data Encryption Algorithm (IDEA) [6]. IDEA seems to be taking the forefront, although it is much too early yet to predict winners. All follow the principles suggested above.

The U.S. Government has decided to promote Clipper, a hardware-based encryption scheme, which it would control, but its attempts to make this an obligatory replacement met with very strong opposition, and Clipper is now likely to be restricted to Department of Defense applications. The number of "unauthorized" schemes in widespread use is also increasing.

The privacy of electronic mail and transactions sent over public networks, particularly the Internet, has become a major issue. The lack of ownership and any standards-setting body for the Internet means that there is unlikely to be any single solution. Again, key management is probably more of the issue than the algorithm itself. The critical question is whether privacy, integrity, or authentication is the main requirement; the best encryption solution will be different for each.

7.8 IDENTIFICATION TESTS

7.8.1 Types of Identification Tests

Frequently, there is a requirement to identify or prove the identity of the person sending the message.

Identity can be established by:

- Something we *own*: A key, a smart card, or other device;
- Something we *know*: A PIN, a password, or the answer to a specific question;
- Something we *are*: A physical characteristic such as a fingerprint or the pattern on the retina of an eye, the way we sign our name, or our overall appearance (comparison with a photograph).

Something we own can be stolen or duplicated, and knowledge can be passed on (inadvertently or deliberately). Much effort has therefore gone into developing reliable tests for physical characteristics: so-called *biometrics*.

The other approach that is frequently used is to combine two types of tests: bank card systems traditionally depend on the combination of something owned (a card) with either something we are (a signature) or something known (a PIN). Computer system access may depend on a smart card and a password.

7.8.2   Biometrics

Many biometrics are considered unacceptable socially (although they may still be used in a closed environment such as a military installation): these include fingerprints and retina measurements (which involve shining a low-power laser into the eye). Thumbprints have been used commercially on quite a large scale (such as the Seville World Fair in 1992) and seem to be gaining acceptability. Palm measurements require somewhat more cumbersome equipment, but appear to work fairly well and are used by the U.S. Immigration Service for regular travelers. Voice patterns have been tried, but people's voices change too much according to conditions.

Signatures are less contentious, but the wide range of variations in signature patterns make them difficult to test automatically (and visually, in fact). The patterns of pressure and speed we use when signing apparently vary less, and systems that measure these, known as *dynamic signature verification* are found to be more reliable.

All of these tests suffer, however, from the problem that they are not absolute: there is never a perfect match with the original data. Some people are falsely rejected, and there is also a risk of impostors being accepted. The tests can generally be tuned in such a way that either one or the other risk is very low, but not both: if the criterion for matching is too strict, then genuine customers may be rejected, while if the criteria are relaxed, the impostors are accepted. The banks have been particularly interested in this problem, since, as we will see later, there are many problems with PINs. They have set as a condition that any biometric used in a payment system must be capable of giving simultaneously a false rejection and false acceptance rate of less than 1 in 10,000. This cannot be met by any current biometric on its own.

Where biometrics can be useful, however, is in giving backup evidence. For example, when a person's identity has already been established in principle (by a card or ID/password combination), the biometric could be used (with fairly lax criteria) to give further evidence of identity.

7.8.3   Zero Knowledge Tests

A similar principle may be used with knowledge tests. Although an impostor may learn the answer to a fixed question, he or she is much less likely to be able to answer a series of questions, particularly if the sequence cannot be predicted. The *zero knowledge* principle is useful here. A zero knowledge test asks enough questions to ascertain that the person knows the secret (is the right person) without actually divulging the answer itself. The level of certainty increases with every question, and for most practical commercial purposes, two or three questions will establish a person's identity beyond reasonable doubt.

References

[1] X3.92, Data Encryption Algorithm, American National Standards Institute, 1975.

[2] Rivest, R., A. Shamir, and L. Adleman, "A Method for Obtaining Digital Signatures and Public-Key Cryptosystems," *Comm. ACM*, Vol. 21, 1978, pp. 120–126.

[3] Shannon, "Communication Theory of Secrecy Systems," *Bell Systems Technical J.*, Vol. 28, 1949.

[4] Shimizu, A., and S. Miyaguchi, "Fast Data Encipherment Algorithm: FEAL," *Advances in Cryptology (Proc. Eurocrypt 1987)*, Springer-Verlag, 1988.

[5] Lai, X., J.L. Massey, and Murphy, "Markov Ciphers and Differential Cryptanalysis," *Advances in Cryptology (Proc. Eurocrypt 1991)*, Springer-Verlag, 1991.

[6] Brown, L., J. Pieprzyk, and J. Seberry, "LOKI—a Cryptographic Primitive for Authentication and Secrecy Applications," *Proc. Auscrypt*, Springer-Verlag, 1990.

Keys and Key Management | 8

8.1 ALGORITHMS AND KEYS

It is now widely accepted that the best combination of system security and reliability will be achieved by using a proven implementation of a strong algorithm. Several of these algorithms have been published as standards. In addition to the DES and RSA algorithms described in the previous chapter, we can mention the Transaction Key systems used in several financial networks.

Prior to the publication of the DES standard, cryptographers generally tried to keep their algorithms secret, although they have always understood that a code breaker may not only know and understand the algorithm, but also have available a large quantity of ciphertext and also probably a good idea of what the data relate to. A code breaker can therefore deduce some of the terms likely to occur in the plaintext. This is really a worst case for the cryptographer, but it is a reminder that it is the key, not the algorithm, that must be kept secret at all costs. The use of a published algorithm makes this reminder even more important.

Although asymmetrical algorithms have an important part to play in secure data communications systems, most data encryption still uses symmetrical schemes, in which the key has to be transmitted from one party to the other. In order for the algorithm to be secure, the key must be long. For the rest of this chapter, we will assume we are using DES (which uses a 56-bit key) or a similar algorithm with a 64- or 128-bit key.

Simple character substitution algorithms can easily be shown to be insecure. But an algorithm that always produces the same result for any given block of 8 bytes can also be cryptanalyzed using the same techniques. Some patterns may recur quite frequently in the plaintext, so it is necessary to change the key frequently if we want our communications to be really secure.

There are some variations on the simple block checking scheme that improve the situation slightly:

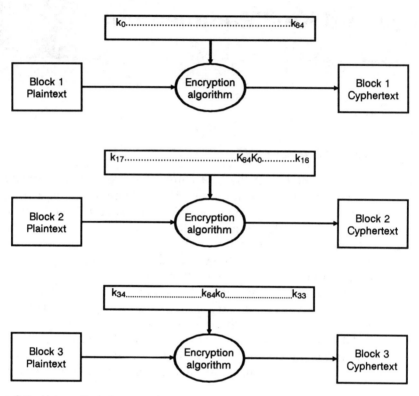

Figure 8.1   Using offsets to vary a key.

- Using a fixed sequence of offsets, as in Figure 8.1, so that the key is rotated every time it is used;
- Using a feedback loop, as in Figure 8.2, so that a few bytes of the previous block are fed into the current block (this method increases the encryption overhead).

Practical commercial implementations balance the need for security with the need for reliability and low overhead. It is, however, necessary to pick an implementation that is appropriate for a given application.

As a general rule, there will always be a need to transmit keys from the host to the slave system within the data transmission, and the design of the encryption scheme must provide for this.

8.2   TYPES OF KEYS

8.2.1   Key-Encrypting and Data-Encrypting Keys

The most common way to send keys in a transmission is to encrypt them using another key, known as the *key-encrypting key*. The key used to encrypt the data

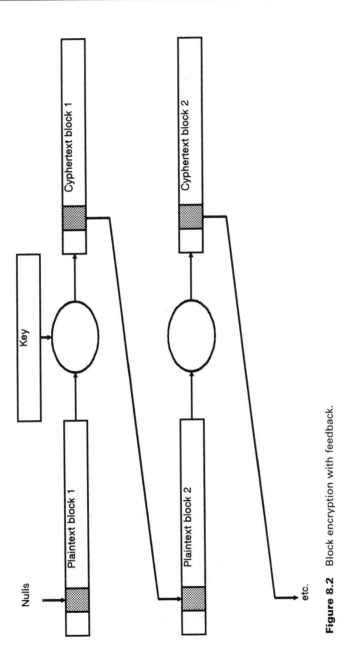

Figure 8.2 Block encryption with feedback.

is then known as the *data-encrypting key*. The data-encrypting key can be chosen truly at random by the host. The key-encrypting key must be known by both ends in advance and is usually kept fixed until changed by a manually initiated action. Each session starts with an initial sequence, including an authentication dialogue (challenge and response). When this has been completed successfully, the host transmits the data-encrypting key, which is encrypted by the key-encrypting key. The data are then transmitted using the data-encrypting key. This key may be kept constant for the whole session, or it can be changed at fixed intervals or after a given number of blocks.

Keys used in this way are often known as *primary* and *session* keys. A session, in this context, can be defined in many ways: it might be a single message, an exchange of messages, a transaction, or the time that the device remains connected. Much will depend on the way the application calls the security software.

For transactional software, a slightly different technique is often applied. At the end of each transaction, and as a part of the transaction, a new key is sent, which will be the key for the next transaction, and it acts as a sequence number and prevents duplicate or missed transactions. This key is known as a *transaction key*; to use it, you need to start the sequence with a *seed key*, known by both the host and the terminal in advance of the first transaction.

More sophisticated schemes store a stack of key-encrypting keys, rather than just one. They are used in succession, with a pointer advancing one after every use. This means that no key is ever used twice in a row.

8.2.2   Multipart Keys

As we mentioned earlier, keys need to be long to be secure. But transmitting long keys can add considerably to the transmission overhead, and we should try to avoid this where possible.

One way of avoiding it is to use two-part keys: one part of the key is kept constant while the other is changed. The constant part is usually called a *customer key*, since it is unique to one customer. It is associated with the primary key, which may now be the same for a whole group of customers. The customer key is added (possibly in scrambled form) to any key used in the system.

8.2.3   Asymmetrical Primary Keys

When there is a need both to authenticate users and to encrypt data, a combination of public key and symmetrical algorithms may be used. The initial dialogue contains a challenge and response (see Figure 8.3). Once the host has authenticated the remote user (by validating his or her signature with the remote user's public key), it generates a session key and transmits it under cover of the

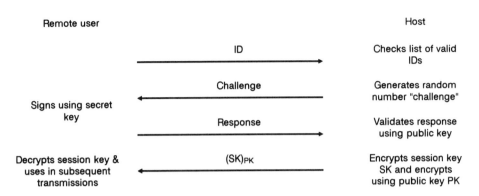

Figure 8.3 Initial dialogue using asymmetrical keys.

remote user's public key. The user validates the signature block and then decrypts the message containing the key.

The only problem with this method is that it requires the use of the relatively complex and slow decryption process for the asymmetrical algorithm. But it does avoid the need for any symmetrical keys to be distributed and stored anywhere in clear (i.e., unencrypted): each end of the link need only store its own secret key.

8.3 KEY MANAGEMENT

It will be clear from the above discussion that good key management is an essential part of any secure system. Not only will the system only be as secure as its key storage and distribution system, but with a poor key management system, recovery from errors can be virtually impossible.

One of the most damaging things an intruder can do is to activate all the alarm mechanisms, which will almost certainly require central system operator intervention. In the worst case, it may cause keys to be destroyed altogether, making it impossible to recover the data at all. Secure storage of keys is therefore an essential requirement. In an environment where the host computer can be regarded as very secure, copies of all permanent or semipermanent keys may be held on the host, encrypted by an overall master key. The master key should not normally be stored on the system intact, however. It should probably be held on a smart card, protected by a PIN known only to one or two people. (It may be dangerous if only one person knows the PIN: what if he or she dies?)

Very often, the security of the host system is just what we are trying to protect. In this case, a further level of key protection is desirable, so that both host and remote user have to cooperate to obtain or use the key. Keys, whether permanent keys such as those discussed above or the short-lived keys used in data transmission, must be transmitted securely. Messages in which a new key is being transmitted should ideally incorporate redundancy for error correction rather than rely on repeat transmissions, unless the redundancy breaches the separation of data link layer and application layer functions in a "true" OSI system.

ISO 8732 (which corresponds to ANSI X9.17) describes a multilayer key management scheme used in interbank communications. It can also be used in many other applications for communicating between trusted hosts and other hosts and terminals, and is the basis of several key management software products.

ISO 8732 defines a set of *cryptographic service messages*:

RSI: request for service initiation (set up new keys);
KSM: key service message (send a key);
RSM: response service message (send authenticated response);
RFS: request for service (initiate key distribution);
RTR: response to requestor (return distributed key);
DSM: disconnect service (erase key);
ESM: error service (report errors);
ERS: error recovery service (recover from errors).

These messages can be used by a terminal, agent, or host, and so allow a lot of freedom in the design of the system architecture. Systems using ISO 8732 can operate on a peer-to-peer, trusted host or key distribution center basis. In a peer-to-peer network, each computer system manages its own keys and authenticates the responses from the other computers. In a trusted host system, the host performs the authentication. In a key distribution center system, the host also issues the data-encrypting keys to all parties.

Much of this section has been concerned with the design criteria for encryption systems and key management software. But this is one area where proprietary packages should be used unless they are completely unsuitable. Not only is it difficult to specify your own software and to find staff able to implement it, but once delivered the software will never be as secure as one written and tested by an outside organization with no knowledge of your systems or probably even your identity.

8.4   FILE STORAGE

One area in which customers may have more freedom of choice is encryption of stored files. The first thing to decide is which files require encryption. Files on disk are often compressed, and this affords a small level of protection to begin with. Full encryption of files does involve a risk (that it may subsequently be difficult or impossible to recover the data) and an overhead in encrypting and decrypting. It should not be applied to all files in a system.

A few operating systems (mostly mainframe or supermini systems) may be regarded as secure enough not to need any further protection. Otherwise, any files that could of themselves be harmful to the company if they came into the wrong hands should be encrypted in some form.

As we discussed in an earlier chapter, there are some simple steps that can reduce the danger. But some files must be subjected to full encryption. In the latter case, a choice of keys and algorithms must be made as discussed in Sections 8.1 and 8.2. Usually, the encrypting and decrypting software will be in the same package, and so a symmetrical key system (probably DES) will be used.

For maximum security, the encryption and key storage will take place in a hardware security module. For many practical systems, however, an acceptable level of security can be obtained using software packages and local key storage.

If file encryption is used, consideration must also be given to the method of extracting the data if the security is compromised (by, for example, a failed hacking attempt). This may be a simple issue of maintaining backups or copies of files, but conventional file "mirroring" is unlikely to achieve the results required, since both copies of the file are likely to be corrupted or both keys lost. It is better if this aspect is handled within the application software, which should be able to write data in different forms to different files without endangering system security.

Hardware Tools

9

9.1 AIMS

We have in previous chapters discussed the various risks to which computer networks are vulnerable. Some of these are best controlled by organizational and personnel controls or by physical procedures, and others by good system design, which should usually be tackled first. But there are many cases where there is no economical alternative to a hardware or software technological tool.

Hardware tools are the most effective for controlling access by people outside the organization. When it is necessary to be in possession of a device to log on, for example, these tools can be used to prevent hackers or former employees from gaining access.

Hardware is usually faster than software for a given task, such as encryption; it is therefore more suitable when a task needs to be performed within a fixed period of time, which is often the case in data communications.

In many applications, hardware also has the advantage of providing a visual check that can be understood by nontechnical staff, including security personnel. And last but not least, theft of a piece of hardware is without question an offense; this is not always the case for software, where theft is also more difficult to prove.

9.2 INTEGRATED CIRCUITS

Several companies have produced integrated circuits (chips) with specific security functions:

- Encryption chips for standard algorithms, notably the DES and RSA algorithms described in the previous chapter. Some of these chips simply comprise a microprocessor with memory and firmware for this specific task, but more specialized chips are hard wired for the DES algorithm.

- Application-specific integrated circuits (ASIC) for more complex crypto-graphic tasks, including secure key management and storage. An ASIC may form a *cryptographic kernel* for a custom-built system.
- Secure bootstrap programmable read-only memory (PROM) for PCs and similar systems. This kind of PROM replaces the standard PROM used when the PC is first switched on. With a secure PROM, the user must go through a password sequence (possibly involving a smart card or other external device) before access to the system is granted. It should be noted that this method is only really secure if the PROM also controls the en-cryption of the data on the disk; otherwise it can be replaced with a standard PROM to give access to the system and its data.

9.3 DONGLES

Dongles are pieces of hardware that are interrogated by the software before it will run. Most of them consist of a small programmable array concealing an ID number, and they are frequently plugged into a parallel port (see Figure 9.1).

Dongles were originally developed and are still mainly used for controlling software circulation. They prevent purchasers of a piece of software from using it on multiple installations at one time, without actually preventing them from copying it, which is regarded as normal good practice. But a dongle is also a good and simple way of preventing many forms of unauthorized access. Particularly useful are those dongles that do not need to be plugged into the system to work, but use radio frequency waves to communicate with an antenna attached to the PC or terminal.

9.4 SMART CARDS

The next stage up from the dongle is the smart card. This is a plastic card of the same size and shape as a credit card. Within the 0.76-mm thickness of the card, however, is an integrated circuit. Such cards are now very widely used in Europe and in Japan as telephone cards, and most countries are now considering their use as bank cards. But they can also be used as travel passes or admission tickets, and they can hold huge amounts of data, such as a person's medical history.

Some integrated circuit cards are not truly "smart," in that they do not contain a microprocessor and software, but only memory or wired logic. This is the case, for example, with most telephone cards. For security applications, however, the more powerful chips are likely to be used, and these are most definitely "smart."

Most smart cards conform to the ISO 7812 and 7816 standards. The former sets out the shape and size of the card and its durability requirements, which are the same as for a magnetic stripe card. The latter defines the special features

Figure 9.1 Dongles.

of the smart card. Part 3 of ISO 7816, probably the most important part, defines protocols for various functions typically required by bank cards. It has, however, been up to the banking industry, particularly the French GIE Carte Bancaire, followed by Visa and Eurocard/MasterCard, to define how these functions are to be used. This international standardization within the banking industry has taken longer than most would have hoped, and it still leaves other users of smart cards to some extent short of a full set of standards.

There is also an intellectual property problem surrounding smart cards. There were two parallel and independent development paths to the smart card, one in France and the other in Japan. The French inventor in particular, Roland Moréno, has pursued a vigorous policy of patenting and licensing which has deterred many of the more innovative systems companies from making use of the technology. The first patents start to expire in 1995, and this may lead to an expansion of the market.

Smart cards offer a number of important advantages to the security system developer. Not only are they much more difficult to copy than any passive hardware device (or a magnetic card), but their much larger memory size and built-in microprocessor mean that they are able to perform several other tasks in addition to the simple identification role of the dongle.

In Figure 9.2 we show the main components of a typical smart card. Power control is very important: many of the early French bank cards were destroyed by a faulty terminal design. The power is supplied by the reader, and the chip itself powers up when inserted into the reader.

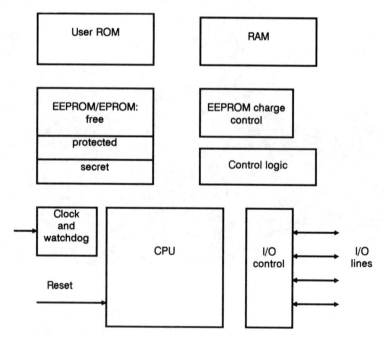

Figure 9.2 Smart card block diagram.

The microprocessor is the engine of the chip. It is supported by an operating system, known as the *mask*, which is stored in read-only memory (ROM). Most of the built-in functions are contained within the mask. The rest of the memory is divided into electrically erasable read-only memory (EEPROM or E²PROM) and random access memory (RAM). Other programs can be loaded into the EEPROM, while the RAM is used for data storage. The RAM may be further divided into more areas which are only accessible to specific functions.

Many of the more powerful smart card chips have built-in DES encryption, and some also feature the RSA signature/verification function. Typical performance for DES on a smart card is around 500 μs for a 64-bit DES block (allowing serial communication at 100 kbps), or 400 ms to sign a 512-bit signature using the RSA algorithm.

Another important function often built into a smart card chip is PIN verification. Once a PIN is stored in the card, a fuse can be blown and that PIN is not accessible to any outside function. The card, however, will only give access to certain parts of its memory after it has been supplied with a correct PIN. When the card is removed from the reader (powered off), the memory is protected again. This function is used by the banking industry to protect against the use of stolen cards, and in communications security it can be used to protect a master key or sequence number.

The smart card is very good at working alongside other techniques. It is relatively simple to carry around and use, and it provides a very effective way of storing a key or other important data away from the main system. In the few cases where the format of the card itself is not convenient, other shapes, such as token or key, can be provided with the same functions. It is not even necessary to have electrical contacts; contactless smart cards will operate simply by being placed close to the reader.

Smart cards are already used for security functions in money transmission, bank cards, and in building and computer network access control. Subject to the resolution of the standards issue (which is not likely to happen overnight), a single card could be used for all these functions. With increased data storage, smart cards could incorporate a biometric test as well. The cards themselves are sometimes thought expensive at $3 to $6, but readers are very inexpensive (really just a set of contacts), and smart card systems are very cheap to run. Smart cards should at least be considered for access rights and encryption key management in many medium- and high-level computer networks.

9.5 KEY GENERATORS

Many secure computer access control systems make use of a portable token or programmable calculator, which are likely to use one of two techniques (or may combine the two):

- Time-based pseudorandom number generation: A new key is generated every 30 or 60 sec. These systems must be kept synchronized with the host system to which they are linked. (A small buffer of five or six keys is usually kept in the host in case the token is slightly out of synchronization.)
- Challenge and response: The host system generates a challenge (a sequence of characters). This is entered into the calculator, which uses internal parameters (a unique key, a separately entered PIN, and possibly a sequence number) to form an encrypted response that is entered into the computer or terminal. A standard specification for this process is given in ANSI X9.26 (Secure Sign-On Standard), which offers a simply implemented password system many times more secure than a normal ID/password combination.

Some tokens can also generate or verify MACs, but this is not usually necessary, since it is the key, not the algorithm itself, that needs to be kept secret.

For very-high-performance communications systems, special key generators known as *guns* are used to generate a pseudorandom sequence of keys. These allow a data stream to be continuously encrypted. Stream encryption using a separate gun must use an algorithm that recovers quickly from errors (this is known as *nonpropagation of errors*). Military and government systems often use optical fill guns, in which there is no direct connection between the key generator and the encryption system.

9.6 INLINE ENCRYPTORS AND ENCRYPTING MODEMS

More likely to be within the range of the normal commercial computer systems with which we are mostly concerned are the many devices offering inline encryption and key management. They are available in three common formats: asynchronous, synchronous, and packet. The first two are used for directly connected serial links, while packet encryptors are designed for use with X.25 packet-switching services. Simple single-line encryptors start at around $1,500.

An inline encryptor will use some external device (a smart card or other token) for initialization and primary key storage. Following an initial exchange with the host or with its counterpart at the other end of the link, it then maintains its own session or message keys. The smart card is used directly only in the setup phase; thereafter, all encryption is performed by the faster hardware in the encryptor. If the device detects the removal of the smart card during use or any long period of inactivity, it will drop the connection.

Some inline encryptors, such as Racal's Datacryptor 64HS, can operate up to the 2-Mbps speeds of the main E1 local exchange circuits. These encryptors will cost up to $10,000. An inline encryptor will often be used in conjunction with a modem for remote access to a computer system.

Other manufacturers have developed devices that combine the two functions. Again, the general principle is to initialize the hardware using a smart card or other external device, and then to use a built-in DES algorithm for encryption at the transmission speed (up to 64 kbps). The initial exchange may include an authentication dialogue using RSA or another suitable technique.

Modems working at speeds higher than V.22bis (2,400 bps) use a range of compression and phase-shifting effects that make them intrinsically more secure. Error detection and correction through retransmission are built into their protocols, and each pair of communicating modems regularly exchanges control information, so that any intrusion or monitoring would require special equipment. The speed of transmission is continuously adapted to the quality of the communications link, which also helps the reliability of the communications. There is currently some confusion as to the implementation of the high-speed V.34 standard, and vendor-specific standards or "interim" standards should be avoided if maximum reliability is required.

9.7 PC SECURITY MODULES

PC security modules are boards that can be mounted in a PC to provide security functions. Most of these are full-size boards suitable for desktop PCs, but there are some limited-function boards available for laptop computers, which are often considered to be at the highest risk. It is only a matter of time before a full-function security module is available in PCMCIA (Personal Computer Memory Card International Association) format (which is the rather thick credit card size that is becoming the accepted standard for laptop computer add-in cards).

PC security modules are able to provide many of the functions described above, including secure boot control, file encryption on disk, fast DES and RSA functions, MAC generation and validation, and built-in key management. The module itself is usually fully enclosed in a tamper-proof housing, which ensures that any critical data will be destroyed if someone tries to break into it.

The security module may provide other security services that really belong in the operating system, but, as we have seen, a PC is not a secure environment and few PC operating systems provide adequate security. The functions provided by the module will include log-on and access rights control and protection of areas on disk.

Modules of this type may be implemented using digital signal processors (DSP), which are specialized microprocessors optimized for high-speed mathematical functions. Critical functions such as DES encryption may be performed in hardware, allowing the security module to operate at speeds much higher than could be achieved by the PC's internal processor, and hence to operate without degrading the performance of the PC.

The cost of a PC security module, currently around $1,000, is likely to rise rather than fall as greater security functions are demanded of them.

9.8   SECURITY FRONT-END PROCESSORS

Several manufacturers now have available security front-end processors (FEP) for PCs and minicomputers (modern mainframes have their own specialized FEPs). The FEP sits between the host computer and any communications link (such as a telephone line or terminal interface). It handles all the access control functions, including user authentication and, where necessary, message authentication in hardware. IDs and passwords or other secure sign-on procedures are implemented by the FEP, so that the chances of an unauthorized user gaining access to the host computer itself are very slim.

Another form of protection implemented by many of these FEPs is dial-back. When a user calls in to the host, the call is intercepted by the FEP, which requests an ID from the calling terminal and then disconnects. It then dials the user back on the number registered for that user ID. This process can also be used for X.25 and other digital connections. For added security, this dial-back process can take place silently—that is, the normal modem answer tones are suppressed and the FEP does not issue commands such as "Enter ID:" or even the ENQ character. The remote terminal simply waits for a connection to be established and then sends its ID string, which is not echoed. This can prevent hackers or other casual callers from knowing that they have connected to a computer system.

Other nonstandard modem facilities can also be provided, although they will usually require all legitimate users to have similar equipment, which restricts the usefulness of the technique. It is worth remembering that the aim is to have reliable communications as well as high security.

Security FEPs have a wide range of prices and functions; a typical range of prices for off-the-shelf items would be $5,000 to $20,000.

9.9   TEMPEST HARDWARE

Tempest is a standard for protecting hardware against electromagnetic radiation. It is principally used in military applications to prevent eavesdropping and to prevent damage to equipment from high-powered radio frequency transmissions or the fringe effects of a nuclear explosion. Another use is in aerospace, where the concern is to protect the equipment against damage or malfunction as a result of the higher levels of background radiation prevalent in the upper atmosphere and in space.

Tempest hardware and associated cables and connections are heavily shielded; that is, they are surrounded by an electrically continuous metal case or shroud, which prevents electromagnetic radiation from entering or leaving. The higher the frequency, the easier it is for radiation to enter and affect other equipment (although the range is likely to be shorter). With increasing levels of very-high-frequency electromagnetic "pollution" in some environments (e.g., close to satellite dishes), there is a good argument for using Tempest hardware in some critical commercial environments.

The fact that radiation cannot escape from the equipment also makes it less prone to eavesdropping. Even a PC radiates energy, and the frequencies and levels used in conventional (cathode-ray tube) monitors are certainly suitable for monitoring. But the technique requires a substantial amount of signal processing, and it is particularly difficult to regenerate the synchronization signals, so most stories of data being monitored by detecting radiation from outside the PC are probably more theoretical than practical. Nevertheless, practical eavesdroppers have been built and demonstrated, and the use of Tempest hardware should be considered in critical applications.

Tempest designs are of high quality (and expensive) and should be capable of withstanding most forms of direct attack. They should certainly be considered where keys are to be transmitted or stored in clear.

9.10 SPREAD SPECTRUM

Another technology that is finding its way from the military into the commercial sector is the spread spectrum technique, which is applicable to all types of radio frequency systems. Its main use in commercial communications has been in conjunction with satellite systems.

Spread spectrum involves using a bandwidth (frequency range) much greater than that required for the signal itself. One technique is frequency hopping: the carrier frequency onto which the voice or data are modulated moves around within the broader range, either according to a fixed sequence or by signals encrypted within the data stream itself. The decision to change frequency can be based on time or error rates, making the transmission more reliable and better able to resist interference or deliberate jamming.

More sophisticated spread spectrum techniques rely on code-division multiplexing and code-division multiple access. Data streams from a single source are extended into highly redundant code words, which can be picked out, using correlation techniques, from the sum of all the transmissions in the same band. This sum appears to any outsider as white noise—completely random signals—so that there is complete confidentiality of traffic flow as well as of the data. To make sense of the data, the recipient must know the code words of each channel.

Spread spectrum communications are now very widely used for private mobile radio, particularly by police and military users. They are also used for satellite communications, where they avoid unauthorized reception of the signals and minimize interference. The hardware can now be incorporated in almost any radio transmitter/receiver at very little extra cost.

Software Tools   **10**

10.1   AIMS

In contrast with hardware security tools, software security is invisible, can be copied but not stolen, and can operate anywhere across a network. For a given function, software is usually cheaper to implement than hardware and can easily be adapted to new situations or upgraded if security needs to be strengthened. It is dangerous, however, to assume that a simple software implementation will be as secure as a comparable piece of hardware: a good methodology is required to produce good software.

Many security experts are very nervous about software, simply because they are not software experts. Of more concern is the fact that many software experts are nervous about security, because they know that it is the nature of software, and the way it is constructed, that in any significant system there will always be errors. Errors may cause the software to stop functioning under some circumstances or may "misbehave," causing an unexpected or unintended result which may or may not be noticed at the time. Programmers can also leave deliberate errors or even hooks in the software, which can be exploited by other programs or accomplices using the software. Most software errors, however, are the result of undefined conditions: situations that the author of the software had not foreseen. This can be the result of bad specification, bad methodology, or the choice of an inappropriate tool, package, or language.

Older computer languages, including the widely used C language, are much more prone to this than newer languages. Fourth-generation languages very rarely encounter such conditions, which can be virtually eliminated if suitable formal software design methodologies are used. We discuss these later in this chapter.

This is not to say that all formally designed and correctly programmed software programs perform the intended function! The way to verify that software is working properly is through testing. There must be a test specification associated with every level of specification. In many situations, a test environment or test protocol will need to be created and may need to be run every time any modification is made. The rule of thumb is that testing should last between one and two times as long as the specification and design at every stage.

Data communications software is particularly difficult to design, since timing can often be critical and the software must cope with a wide variety of different conditions. A methodology that recognizes states and transitions, rather than a sequential logic, is often preferred.

In this chapter we look at the various forms of software tools that can be used to enhance the security of a network. Such tools may work at the local system level (a PC or terminal on a LAN), the function or program level, the file level, or across a network (see Figure 10.1).

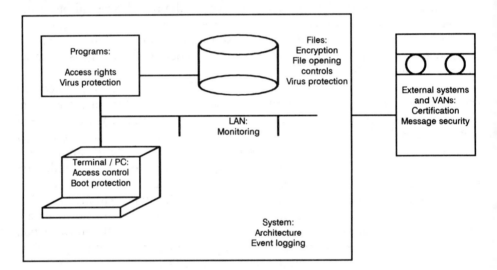

Figure 10.1 Objects protected by software.

10.2 ACCESS CONTROL

10.2.1 Password Control

Most computer users are familiar with basic access control systems. When a user logs on by giving a name or identifier (an ID), he or she is required to give a password, without which access is denied. With simple systems, once a user has logged on, he or she has access to all the facilities of the system. More commonly, each user is allocated certain rights, such as the right to run specific programs or to access specific files or directories. He or she may have the right to load new software, but will still be denied access to functions the systems regards as privileged or belonging to another user. Access rights are considered in more detail in the next chapter.

As a next stage, many packages are available that give the system manager a further level of control over the log-on process, such as limiting certain users to specific machines or times of day. Users may also be recognized and given a specific startup menu from which they can only access certain functions. These packages may work in conjunction with a hardware tool: a magnetic card, dongle, or even smart card.

The weakest part of most password systems is the people who use them. Any system administrator knows how difficult it is to persuade users to change their passwords frequently, not to use their husband's or girlfriend's name, and not to write their password down beside the terminal. To be really effective, however, passwords must be long, meaningless strings of characters. These are in fact very difficult for most people to memorize, and so passwords on their own are really not recommended for any security-conscious installation.

10.2.2 Boot Protection

A further step must be taken if any standard hardware and operating system (particularly a PC system) is to be protected using a control which is only used when starting a session or turning the system on. We must prevent the user from starting the system using a floppy disk or other method, and then changing to the "normal" operating system having bypassed the initial password session.

Since most hardware has specific provision to allow the user to start the system by an alternative method (e.g., to diagnose or fix a fault), the route taken to provide the extra security is to prevent access to the disk or disks if the system has been started (*bootstrapped*, or just *booted*) by another method. This is difficult to achieve reliably without hardware changes to the system; programs have been written to overcome most of the software-only protection.

If the device (whether software only or hardware and software) is completely secure, then the other danger is that someone may forget a password, leave the company, or be off sick. Or a fault in the protection mechanism may leave all the data in the system inaccessible. Most such systems therefore have a bypass feature in the form of a supervisor disk or master password. The bypass feature is often the weakest link in the security chain.

Once we have ensured that the user cannot boot into the operating system stored on the disk we want to protect, we must also ensure that no other operating system (a floppy-based version, for example, or the remote operating system in the case of a remote terminal) can read the disk either. And we must ensure that any attempt to exit or "shell out" to the operating system from within an application is blocked or leads directly to a reboot. There are many potential escape routes of this type, and an experienced hacker will be aware of most of them.

10.2.3 Authentication

A much more secure route for initial identification of a user to a system is *enhanced access control*. Under this system, the host system issues a challenge to which the user must respond correctly. The challenge is normally a random number, and the response will usually be a function of the challenge, the user's password or PIN, and possibly also the date and time. The user needs some computing power to generate a correct response; this power may take the form of a special calculator, a smart card, a program in an intelligent terminal (such as a PC), or a function contained in a security module or special hardware.

In the case of one computer system logging onto another, the function can be fairly complex: a hashing algorithm followed by RSA encryption using the user's private key would be normal and would provide the basis for further access rights management, key management, and encryption functions. Such functions are available as an option for many standard mainframe systems. They completely defeat intruders who depend on trapping ID/password combinations or on trying large numbers of likely combinations.

10.3 FILE PROTECTION

The next technique is applied to individual files rather than to the whole system. Although all files in a system may be encrypted, the encryption will add significantly to the processing overhead. The most powerful packages allow any single file or directory to be comprehensively protected, which is more cumbersome to set up, but safer than systems that offer a lower level of protection to all files.

Three broad techniques are used:

- Encryption of the whole file, requiring it to be completely decrypted before use. This has the danger that there is a period when the file is available in clear. A different key is used each time the file is written away. The management of the list of keys becomes a very critical task.
- Encryption of the complete file in blocks, so that each block must be decrypted before it can be read. This is suitable for certain types of files only, and is very cumbersome if frequent key changes are required. It is most appropriate when integrated into application software; at least one vendor provides this type of software in the form of "C" modules.
- Modifications to the operating system so that opening a file is dependent on a password or other initial security procedure. The data may be held in clear. This is a neat solution, but it is not suitable for high-security environments.

All these systems have their limitations, and so it is probably best for individual file protection to be determined by the application software, using modules such as those described in the second example above, or specially designed encoding and authorization techniques. This is, though, not an option for users of standard packages (databases, spreadsheets, and word processors), who must rely on their operating systems and other standard packages for protection.

10.4 VIRUS PROTECTION

A major concern for managers of computer networks is the possibility of infection by computer viruses. A virus is a program that can duplicate itself; its effects may be harmless (some are never seen but give their authors the satisfaction of knowing they are there), irritating (such as screen messages), or devastating (some programs can overwrite file access tables or disk boot sectors).

Early viruses were mostly "Trojan horses"—small programs sent in to make a future hacking attempt easier, usually by copying or recording files of IDs and passwords. The ability to replicate themselves was a bonus rather than a necessity.

Relatively benign examples include the Cookie Monster, which "eats" characters off the page until it is left blank; the Christma virus, which displays a Christmas tree; and the DRIP virus, which causes all the characters to fall to the bottom of the screen. None of these actually destroy data, although they can waste a lot of time for those who have to eradicate them.

More problematic are viruses that consume resources without actually doing anything. The Morris virus, which infected the ARPANet (the predecessor of the Internet) in 1988, just sat in memory doing complex arithmetic once it had been loaded by a user accessing a mailbox.

The most vicious viruses, however, are those that deliberately destroy data, often the boot sector of a disk, which makes recovery very difficult. A recent development is the polymorphic virus, such as Dark Avenger, which changes its form with every duplication, thus avoiding virus detection systems that rely on a characteristic signature of a known virus.

Viruses are almost always introduced to systems through unauthorized software, although there have been cases of viruses appearing in legitimate boxed and sealed software from smaller companies. There are three main techniques for protecting against virus infection, and they are usually used in combination:

- As each new virus becomes known, some characteristic of its code is isolated and stored in a library. A program can then check a whole disk or every executable (program) file for these characteristic signatures. It is much safer if the whole disk is checked, since a virus can actually lie dormant until it is "hit" by a program. These programs depend for their effectiveness on being updated frequently as new viruses become known. Running them can take a long time, but should be an infrequent task; they only detect a virus infection after it has happened.

- Since viruses are normally spread by being run, they must infect a program file. The second form of virus detection relies on checking all program files to ensure that they have not changed in any way since they were originally loaded or checked. This type of check is run frequently and should detect a virus infection before it can do significant harm.
- To prevent virus infections, all possible sources of new executable programs must be checked before they are run. Many users should simply be prohibited from introducing new files of any kind. A high proportion of virus infection is introduced by temporary staff or by engineers working on the system, and these situations should be particularly tightly controlled. Several packages are available to monitor and control all sources of new software, but, again, it is important to ensure that any required bypass routes are strictly controlled.

Virus protection is now becoming commonplace, and all three forms of protection should be used on any installation with wide access. Figure 10.2 shows the minimum requirements for virus protection.

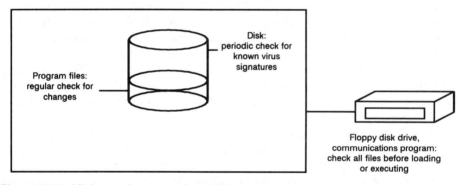

Figure 10.2   Minimum virus protection requirements.

10.5   OPERATING SYSTEMS

10.5.1   Standard Features

The most critical software element is the operating system. The choice of an operating system is always connected with many other decisions, so that often security will be one of the last factors considered. It is nonetheless important to be aware of the extent to which the operating system can influence security.

Most operating systems would probably be regarded as neutral from a security point of view: they do not of themselves do anything to enhance or detract from the security of the system. There are, however, many factors that do differentiate even relatively normal operating systems. As we discussed in Section 5.2.1, these factors include user-friendliness, stability, security services, and interprocess protection. Even the most basic operating system today (it was not always so) has some form of file protection: files can be hidden or marked as read-only. But this protection rarely extends to communications paths. A neutral operating system is unlikely to be sufficiently secure for use in a data communications environment.

The lowest level of protection necessary will be some form of *enhanced security services*, which users or application programs can call upon to protect data and communications processes. These services range from password protection on files to full encryption available on both files and communications paths.

Most multiuser operating systems incorporate some form of *interprocess protection*; that is, each process is allocated certain resources, so that it should be impossible for one process to affect another except by communicating through the operating system.

For higher security, interprocess protection is combined with *interuser protection*. This type of protection is inherent in the system and will be independent of password controls or the levels of access accorded to individual users. Even two low-security users are fully protected from one another.

Most multiuser systems do, on the other hand, have an *access control* system, including some concept of user rights, which are discussed further in the next chapter.

10.5.2 Client-Server Systems

Security-aware systems will include the concept of a *security level*. Every process or class of data has a security level associated with it, and users whose security level is not adequate will not be allowed access to the program or data. This reduces the danger of a user being able to find a route into a higher security process, and can often be important in the context of modern *client-server* systems. In these systems, many processes are run by so-called *agents*, which may be other processes. A secure process can therefore be called by a less secure process, and by this route—a classic hacker's route—a user may gain access to inappropriate data.

10.5.3 Monitoring

One often underestimated feature of an operating system—important for good management as well as security—is the level of *monitoring* and *logging* it provides. With traditional mainframe systems, there would usually be a full-time operator or operators, whose screen would continuously report such activities as users logging on and processes starting and completing. Often there is just too much activity for this to be useful, and some form of exception reporting may be preferable. On today's smaller systems and PC networks, if there is a system supervisor logged into the network, he or she can receive the same information.

Again, he or she will only make use of it if there is some form of exception reporting. If there is no supervisor, or if he or she is not logged in (at night, for example), some action should still be taken on exceptional conditions: perhaps the user who caused it should be locked out immediately or another screen should receive a message. In all events, a wide variety of system events should be logged to a file so that they can be traced back afterwards in the event of an incident.

10.5.4 Security Architectures

Only a very small number of operating systems can really be said to incorporate a security architecture; that is, they are designed from the outset with security in mind rather than having security features added.

Interprocess and interuser protection are the first essentials. They are features of most supermini and mainframe operating systems. A wide range of security services available to programs will also be found in these systems.

The OSI standards include a framework for a security architecture (ISO 7498.2). The main contribution of this standard is to define the main security services and allocate them to layers in the OSI Reference Model (see Section 5.2.6). This allocation is illustrated in Table 10.1.

Table 10.1
Allocation of Security Services to OSI Layers

	Layer						
	1	2 Data Link	3 Net-work	4 Trans-port	5 Session	6 Presen-tation	7 Applica-tion
Service	Physical						
Peer entity authentication			Y	Y			Y
Data origin authentication			Y	Y			Y
Access control service			Y	Y			Y
Connection confidentiality	Y	Y	Y	Y		Y	Y
Connectionless confidentiality		Y	Y	Y		Y	Y
Selective field confidentiality						Y	Y
Traffic flow confidentiality	Y		Y				Y
Connection integrity with recovery				Y			Y
Connection integrity without recovery			Y	Y			Y
Selective field connection integrity							Y
Connectionless integrity			Y	Y			Y
Selective field connectionless integrity							Y
Nonrepudiation of origin							Y
Nonrepudiation of delivery							Y

But there is a conflict between the needs of security (in the sense of protection from deliberate acts) and ease of use. An operating system designed first and foremost for security will never be as easy to use, and certainly not as flexible, as one designed for mainstream applications. Most readers of this book will have to accept the features of a mainstream operating system and make the best use of them.

10.6 NETWORK-SPECIFIC TOOLS

Using external computer networks (VANs) can, as we have seen, be a source of security risks in that they can permit outsiders to gain access to a computer system. But many VANs in fact add to the security of the network, primarily by making it more reliable than any single link could be. Most VANs use multiple paths so that no single equipment failure, either within the network or in public circuits, can prevent a link from being established. They use high-quality digital links, which have much lower error rates than analog links.

Where all external access to a computer system is through a private network, the network itself can also provide security features to complement those in the access control system.

10.6.1 Network Security Services

Networks can provide external authentication and certification functions; in other words, user A can ask the network security server whether this message from user B is legitimate, or user B can attach to his or her message a "certificate of genuineness" from the network. These services usually follow the subdivision of network security services recommended in the OSI Security Architecture, with a number of service messages for key exchange and authentication. This technique is used in several closed user groups (such as the French interbank system) and has also been incorporated into commercial products under the names Kerberos and SESAME.

Other networks may provide less general services aimed at the specific functions required in their application. For example, EDI networks will usually have to provide nonrepudiation services for both senders and receivers of a message. Other networks may provide a high level of integrity through the use of block checks and forward error correction techniques.

10.6.2 Network Message Security

Different tools are available for various applications commonly used on networks. In EDI networks, the text is usually sent in clear (this is a requirement of many EDI services). But the EDIFACT (Electronic Data Interchange for Administration, Commerce, and Transport) standard allows for a signature (probably an RSA-encrypted form of the message CRC, date, and time) that allows the recipient to check the origin and integrity of the message received. This function is included in many EDI software packages.

For electronic mail, the widely accepted standard is the ITU X.400 series and the linked X.500 directory standard. X.500 provides for the storage and distribution of a user's public key, and is used by the Privacy Enhanced Mail (PEM) standard, which is recommended for use on the Internet. PEM is supported by some of the commonly used Internet access managers, but some information providers have preferred proprietary software supported by RSA-based packages such as Pretty Good Privacy (PGP).

For on-line transactions, particularly in the financial arena, further standards are required, which are usually built into the transaction software and are likely to be based on one or another variant of the transaction key principle described in Chapter 8.

10.6.3   Network Monitors

Monitoring and audit functions are very important elements of any security scheme. It is particularly important to monitor and detect failed attempts to log on or access files; repeated attempts may result in success, and failed attempts can leave the system unusable by legitimate users. It is also possible to detect a case in which the security system is preventing a legitimate user from gaining access to the system or a file, perhaps because of a wrong setup parameter or changed password. These cases need to be investigated and corrected.

Most of these functions will be provided by software using a network management protocol such as the Simple Network Management Protocol (SNMP) or its newer and more efficient cousin SNMP2. Most modern network hardware provides direct support for SNMP, but in some environments the use of a hardware analyzer can add some information. Hardware analyzers are themselves potent bugging tools and so need to be protected by good physical security.

In Chapter 5 we talked about the risks associated with network bridges, routers, and gateways. While these risks cannot be eliminated completely, we can at least detect any attempt to tamper with the network or any failure that results in part of the network becoming unavailable. With suitable network-monitoring hardware and software, an operator can be alerted to any unusual situation on the network. Network monitors are recommended for any large network under the control of an identifiable system operator.

10.7   SOFTWARE DESIGN TOOLS

In Chapter 5 we pointed out that good software design involves not only ensuring that the software does exactly what its specification says, but also that its specification meets the end user's requirements. It is particularly difficult to ensure that every possible condition is handled; programmers writing code for a fighter aircraft are not likely to be experts on the weapons or fuel systems, and still less are they likely to be pilots. In a data communications system, the range of possible external events is almost as wide.

To achieve a "correct" system under these circumstances (one that meets its specifications at all levels), a variety of formal programming methods (also known as *structured analysis and design*) have been developed. They cover the specification, design (structuring and coding), verification, and validation stages of the process. The best known examples of these methods are the Jackson Structured Design, Yourdon, and SSADM (Structured System Analysis and Design) techniques. Large programming projects nowadays normally use a structured design methodology, and these methodologies should be considered on secure systems of any size.

For greater security, they should be used in conjunction with a formal notation based on mathematical concepts, which can be translated directly into code. Formal languages must be used in the design of any truly secure system, and their use is a condition of the higher levels of ITSEC certification. Examples of formal notation languages are Z, RAISE, and the ISO specification language LOTOS. These languages are more specialized, although they are frequently used by the designers of safety-critical systems such as nuclear reactor control and railway signaling.

Modern programming languages make increasing use of *objects*. A programming object is not a physical entity, but is something that can be defined fairly precisely in words. It may be a set of data fields or a record, the state of a process, or any transient idea that can be manipulated by other functions or objects.

The use of object-oriented programming languages such as C++ should also help in eliminating undefined conditions and similar programming errors, particularly in systems that are frequently upgraded.

10.8 HACKING TOOLS

Just as software can be used to enhance security, it can also be used to break the security of a system. A system manager should know what tools the would-be intruder is likely to use, and must also ensure that the company's own systems are not being used for hacking. Most hackers start off with fairly simple equipment—a PC and a fast modem. The first essential for a hacker is a good set of *emulators*. These turn the PC into whatever type of terminal the host is expecting to see. Older mainframe systems usually expected a "dumb" terminal with no buffering or special codes; but most terminals today emulate one of the Digital Equipment, IBM, or Tektronix terminals, or consist of special communications software designed for PCs.

File transfer protocols also vary from system to system, but most follow an accepted de facto standard such as YModem or Kermit. Many PC communications packages have all these options, and systems managers should ensure that any such packages are only set up for the systems they will legitimately be used with.

The next essential is an *autodialer.* Almost every modem has an autodialing facility, but some are more versatile than others (the communications package must also allow the user access to the extra commands). The almost universal Hayes command set uses the string "ATDT" (or just "ATD") for dialing.

Many hackers use these commands to write their own software for trying out telephone numbers and recording dialogues. Digital systems (X.25 and ISDN) have their own command strings and procedures that are the equivalent of the "ATDT" commands.

Particularly useful to the hacker are programs that can try many combinations of numbers: *power dialers.* These programs can try many numbers and record any that gave a modem carrier tone. The system manager should be on the lookout for files containing many telephone numbers, the "ATDT" sequence, or long data files associated with any communications program.

Calls that connect and then drop the line can be the first sign of an attempt at intrusion. They are likely to be followed by repeated attempts to log on, many of which will usually fail because of technical incompatibilities.

If allowed to continue, these calls may often succeed in the end, usually because there exists some leaked password or alternative route into the system. This emphasizes the need for strict control of all possible access routes as well as good password discipline and system monitoring.

Access Rights Management   **11**

11.1 Scope

Access rights are at the heart of any computer security scheme. Leaving aside the reliability and accuracy issues, the key question is: Who can do what to which data?

This is a three-dimensional question:

- We must divide the *data* into different sets according to the security requirements of each.
- *Users* must also be subdivided. Users may fall into several groups according to their functions, departments, and positions in the organization.
- The *rights* that can be granted or controlled will depend on the operating system. In most cases, users can be given or denied the right to read or write data, to create new files, or to delete data. This is often shown on a simple table, as in Table 11.1. As we will see later, the division can be much finer than this, and may need to be in some cases.

The role of an access control scheme is to define these rights and to say how they will be controlled. The first level of control is usually physical; then there are the levels of control afforded by the password or log-on system used, together with the operating system. Finally, there are the controls that have to be written specially or incorporated via a hardware or software security package.

A good access control scheme is a precondition for a secure system. It is all too easy to leave everything in the hands of a system administrator, with the common result that users are unable to perform their normal tasks, particularly those dealing with unusual cases, without referring to the system administrator. This one person also has an excessive level of power, which can result in an accidental loss of data or security breach, or a deliberate breach which would be difficult to detect or prevent.

The scheme needs to be thought through and discussed with managers as well as computer staff. It must be written down and included in any system

Table 11.1
Simple Access Rights

Users	Data Types			
	Accounting	Personnel	Stock	Production
Managing director	R	R	R	R
Financial director	RWCD	—	R	R
Bought ledger clerk	RW	—	R	R
Production manager	—	—	RW	RW
Stock controller	—	—	RW	R

Abbreviations: R = read, W = write, C = create, D = delete.

manuals or business procedures. And it must actually reflect the business process rather than an idealized or compartmental view of the business. Before sitting down to design an access control scheme, you really need to have a flowchart of the business process and to understand who needs to know what information.

Access rights may be managed at several levels:

- Physical access control to prevent unauthorized personnel from gaining access to the network at all or to limit access to certain terminals, printers, and communications equipment and lines.
- At the beginning of a session, both by controlling system access through password controls and by controlling users' access to functions and menus. This is a rather coarse set of controls, since they cannot be tailored to the application. But many standard applications do not have any built-in controls. Strictly, the OSI model does not allocate any security services to the Session layer. Password controls should be effected at the application layer, even though their effect is on the session.
- At the Application layer, within the application and by the use of support services (e.g., romote system access) operating at this level. This is the most detailed level of control, but is also usually the most expensive to use, both in terms of design time and overhead when running.

11.2   PHYSICAL ACCESS CONTROL

Although physical access control is the lowest level of control, it remains one of the most important. A good physical control system gives an immediate impression of security and may deter attempts to go further.

It is said that "out of sight is out of mind." Many breaches of security depend on the intruder having access to the system and seeing that there is useful information to be gleaned. Temptation is often a major factor in insider fraud, particularly where lower grade staff are concerned. A small deed leads to a bigger one and so on down a slippery slope.

Physical controls should cover not only staff, but also equipment entering and leaving an office. Passive radio or inductively powered tags can be used where the controls have to be automatic. Special inventory tags, securely glued to the equipment, are activated when they are taken past a sensor.

11.3 SESSION LEVEL ACCESS CONTROL

Password systems were considered in the previous chapter. There are also simple physical controls that can be used for terminals, including keyboard or power locks. One major PC brand now offers an infrared-controlled power switch; it can only be turned on by a user who has the necessary remote control unit.

With the increasing use of remote access and "hot-desking" (desk sharing by staff who only spend part of their time in the office), it is useful if terminals and ports can be configured correctly for each user as a result of the log-on sequence. This again points to the use of a smart card or other hardware-assisted log-on mechanism.

To avoid equipping all PCs with special smart card readers, at least one company has developed a smart card microprocessor in the form of a floppy disk (see Figure 11.1). It is inserted into a 3½-inch floppy disk drive mechanism,

Figure 11.1 Smart disk (reproduced by permission of Smartdisk Security Corporation (U.K.) Ltd.).

where it couples with the head to exchange information, including key bootstrap files without which the system will not function.

11.4   INTERHOST ACCESS

Even when the user at the remote host is validated, the local host should make use of a two-way "challenge and response" validation using a MAC. The remote host may also want validation on its side, resulting in a three-way process. In some situations, it will be important to identify the user at the remote host rather than just the host itself. This is not always possible using the operating system facilities and may have to be performed by the application.

11.5   HIERARCHIES AND WORKGROUPS

11.5.1   Link to Information Flow

An access control scheme must recognize the way information flows in the organization and how processes are initiated. This is why it is important to analyze the business flow diagram before starting work on an access control scheme. Some companies start out with a simple hierarchy: junior staff at the bottom and the chief executive at the top. The fact that the system administrator has to have more access rights than the chief executive shows the flaws in this scheme immediately. Many managers resent the idea that their staff may have more rights than they do on the computer system; but their job is, after all, to manage. Or do they in fact deal with tricky cases or help out when the load is heavy? How do they control or check their staff's work? More realistically, staff may be placed in departments: production staff have access to production plans and stock data, personnel to the staff records and payroll, accounts to the ledgers (and possibly the payroll as well). Within each department, it becomes rather easier to analyze who needs access to each set of data.

Within an application or in a database environment, normal users are usually defined as having no access, read-only access, or read and write access, while only computer staff have the right to create and delete files. Some operating systems include more specific controls, such as controlling the copying of files. With some utilities, such as spreadsheets or word processors, a wider range of people must be able to create files, and access rights must then be defined for their files.

It is therefore normal to base an access control scheme on workgroups. The starting point for this is the department and the functional groups within each department. An individual may fall into several groups that have different levels of rights over different data sets, as in Figure 11.2.

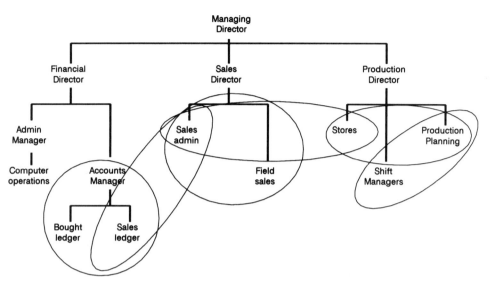

Figure 11.2 Workgroup-based access rights.

As we will see, setting up a comprehensive access rights system is quite a laborious task. Default settings must be used extensively to start off with, which are then modified when the business process chart is examined or when taking into account any particularly sensitive files.

11.5.2 Databases, Files, and Aggregates

Most operating system access control schemes are designed to work at the file level. A user can read or write all or none of the data in a file. With a database structure, however, this may not be appropriate. The goods-receiving department must be able to update the quantity of an item in stock, but possibly not the price or preferred supplier. A product manager may be able to perform a wide range of operations on his products, but should not have access to other product managers' areas. The database system itself must provide the protection at this level in a way, if possible, that is compatible with that provided by the operating system.

There are situations within a database structure in which an application accessing one database needs to consult a second database without having direct access to the second database. In these cases, the application may be granted *delegated* or *proxy* rights, which are a combination of the rights of the person or entity running the first application with some additional rights of its own. For example, managers often have a need for aggregate data without the need to see or manipulate the individual items. This can be achieved by controls on the processes used to extract the aggregate data.

There are other situations in which the requirements of privacy or data protection, for example, demand that individual data are not displayed or manipulated. In some cases, it is even necessary to add a deliberate error, or *noise factor*, to the results in order to avoid the possibility of extracting individual results from the aggregate data.

11.5.3   Testing and Support Modes

Another dimension of access relates to solving both operational and computer problems. Programs and "fixes" run to overcome a specific problem often break the rigid rules of the original system, and this means that they require some special privileges. Where this occurs, the program itself must be subject to more thorough testing, not less, than the original system was, and the security rules for running it must be stronger (without shortcuts).

Computer staff often assume a higher level of rights than they need or fail to change back to a lower level when appropriate. This gives much more scope for their accidental errors to become serious or for others to see new routes through the system protection. Application programmers, for example, rarely need access to the live data files: they should usually work on extracted files, possibly even dummy files, unless the file itself really has no value. Utilities for providing programs with representative dummy data are a useful part of any test suite and should certainly be available to programmers developing communications software.

System testing is a more dangerous situation; there is always a point at which live data have to be fed into the system, and this will usually require a high level of authority. A test mode or test user is a useful tool here. Either of these can be set up when required, but *must* be deleted afterwards. Activity with test users must be treated with great care, and should certainly be explicitly logged.

The most dangerous of all for security is an engineering mode. This often allows a user to perform a wide range of operations at the physical level, thus completely bypassing any software controls. These operations can include copying disks or parts of a disk, writing bit patterns to disk, or monitoring data passing between system components or across a communications link. Engineers can even leave hooks in the system that allow them to view data or data flows when the system reverts to normal operation. Engineers have also been the source of virus infections introduced through test software.

Many engineers are from outside the company, and so they are not known to the staff dealing with them and have not been security checked. Only the most security-conscious organizations insist on evaluating the skill and background of individual engineers. For most companies, the engineer arrives at a stressful time, and any assistance he or she can provide is welcomed without question. Most engineers offer their best efforts in good faith, but it can be worth emphasizing to any engineer that any changes (even temporary) to system tables, standard software, or the use of hardware monitors must be cleared explicitly with the system supervisor, and that any testing modes used must be turned off at the end of his or her visit.

11.5.4 System Administration

The system administrator, supervisor, or manager should be all-powerful. It really is important that one person carries the ultimate responsibility for the security of the system in all respects. Even where a shift system is operated, only one supervisor can be the system manager. He or she must authorize any changes to access rights.

Depending on the size and character of the organization, it may be wise for the system manager to share some of the security responsibility with a senior manager, possibly even the chief executive. Some organizations have a security manager whose role overlaps with that of the system administrator, who may even be responsible to the security manager for computer security issues. These would include the implementation of agreed-on security procedures.

Operations requiring exceptionally high security should require a PIN or other security measure from both parties sharing the responsibility. Initializing seed keys in a transaction key environment may be one such operation.

It is also essential that the system administrator can be trusted thoroughly. It is unlikely that anyone else has the same level of knowledge of the system and surrounding procedures. Nevertheless, the good system manager will, for his or her own protection, ensure that all actions that could compromise security are logged, and that he or she is subject to the same controls as all the other staff.

A weak point in many network systems is the backup procedure. Ensuring that system files are backed up often has to be done by the highest supervisor level, and this can mean that several alternative staff must know the passwords. If possible, such procedures should be controlled by batch processes run automatically so that this problem is avoided. File restoration, particularly of system files, will always require a high level of access rights.

Many operating systems have a "superuser" or supervisor status. Creation of superusers must be a highly controlled process, and it is particularly critical that the creation of these accounts is logged.

11.6 TIME DEPENDENCE

Time functions are a powerful tool if used well. A high proportion of security breaches (both accidental and deliberate) occur outside normal working hours or could be detected as abnormal patterns of use by a suitable analysis program. Staff do often have to work outside normal hours, and many organizations encourage this. However, work performed when physical supervision is less stringent must be more tightly controlled in other ways. It may, for example, be appropriate to restrict users' right to print or copy files (particularly to floppy disk on PC systems or workstations) outside normal hours.

Certain operations, such as financial transactions, should not be carried out during evenings or weekends without a second authorization. Some operating systems, job control systems, or operating system overlays permit a complete change of the access rights pattern outside of normal hours. In other cases, this will have to be handled by the application itself. It is recommended that the controls be put in place in both the application and the operating system where possible.

Job control programs offer another related tool: the ability to ensure that programs are run in a specific order and possibly only at certain times of day.

Time can also be used in another way: to set expiration dates on rights previously granted. This is rarely done, with the exception of a few operating systems that demand that passwords be changed within a set time frame. Some card-based access control schemes use expiration dates on the cards themselves. There is, however, scope for making much more use of expiration dates in access control schemes in general.

11.7   OPERATING SYSTEM ISSUES

The opportunities afforded by a system to carry out the types of control mentioned above depend on what kind of operating system is used. As we discussed in Chapter 10, security will usually only be one factor in the choice of an operating system, and there may (particularly with some hardware) be few alternatives.

11.7.1   Utilities

With some of the more common systems, particularly PC networks, there are utility systems (menu and access control packages) that can be added to the basic operating system to improve the level of control available, and they should be investigated when (as is usually the case) the standard operating system offers an inadequate range of access controls.

11.7.2   Mainframes and Superminis

The main exception to this rule is mainframe and other large-scale systems. A combination such as IBM's VTAM and CICS, or Digital's VAX/VMS, offers a very wide range of security services and access control options. The user must, however, be aware of the limitations of these services and the amount of work involved in setting up a comprehensive scheme for any large-scale network.

VTAM, for example, controls all terminals on most IBM mainframe environments. It offers the facility to force users to log into one application or to prevent them from logging into an application. CICS controls transaction-oriented applications and includes a very comprehensive set of tables for controlling who can run transactions, including time dependencies. The tables themselves, however, are less secure and can usually be modified by systems staff.

IBM's batch job control system JES2 also allows extensive modifications through the use of hooks in the program. The systems development and operator environments, therefore, need to be very tightly controlled when these systems are used, but user operations are well controlled. IBM also has available security products at the application level, such as the Resource Access Control Facility (RACF).

With Digital's VAX/VMS system, password and log-on control is good, but the use of special accounts, especially for remote users, needs to be carefully controlled. VMS features control of access to all *objects* (e.g., files, programs, devices, communications links) through seven general categories and object-specific *access control lists*, although excessive use of the latter can result in very high system overheads.

11.7.3 PC Networks

In the PC world, most of the DOS-based operating systems, including the standard Windows overlay and the popular network operating systems, have relatively low security, particularly regarding protection from faults and loopholes in hardware and software. The newer Windows NT system offers a much higher level of interprocess protection and more access control flexibility, but it is also more cumbersome to set up.

11.8 MANAGEMENT AND SUPERVISION

Earlier in this chapter we mentioned the problems associated with hierarchical access control systems. It is often necessary to address directly the question of management control and supervision when designing an access control scheme, because if a manager must always have at least the same rights as the person managed, then many people will have the right to perform tasks that they should never perform, and the chief executive must have total power over the system. This situation is unlikely to be desirable, and it reduces the security of the system considerably.

Nearly every application should include a management control or log of some form, and the manager should be able to access it. This log will also form part of an audit trail, which is essential in any security-minded installation. A system auditor must consider how he or she would be able to trace the history of an operation, and where sensitive data or financial transactions are concerned, the audit trail would be detailed.

Exception reporting should also be built into most applications to assist in supervision.

11.9   DATA PROTECTION ISSUES

Data protection legislation and attitudes can also have a bearing on the way an access control scheme is implemented. There are wide differences between countries in these areas, and companies operating internationally must be particularly aware of them. In some countries, such as France, the principle of data protection centers around privacy: access to personal records should be minimized, and organizations are generally discouraged or prohibited from storing more personal data (any data that can be traced back to an individual) than is strictly necessary for their activities. In other countries, such as Sweden and Britain, personal data that do not need to be confidential are protected by being made open so that their accuracy can be checked. Some countries exempt public authorities from data protection legislation, while others have no such legislation at all.

In general, however, personal data represent a special class of data. It is often necessary to provide for access to groups or aggregates of the data without granting the right to gain access to individual records. As we mentioned earlier, it is sometimes necessary to suppress output for very small groups or to introduce errors in order to protect individual personal privacy.

Types of Networks

12.1 NETWORK STANDARDS

The nature of the risk in a computer network or data communications system and the types of protection that can be applied vary greatly according to the type of network. Up to the beginning of the 1980s, networks were usually fairly homogeneous, and network standards were dominated by the standards of the major hardware manufacturers, IBM's SNA (System Network Architecture) and Digital's DECNet and VAXNet, partly because it was in practice very difficult to set up a network any other way.

With the gradual, often grudging acceptance of the OSI model by the major hardware and software companies, this situation has changed completely. At the same time, many more technological possibilities have opened up using the same model and raising the need for further standards. The main network standards exist at layers 2 to 5 of the OSI model: layers 2 and 3 in hardware, and 4 and 5 in a network operating system.

12.1.1 LAN Hardware Standards

The main hardware standards for connecting PCs and other computing equipment in a LAN are Ethernet, Token Ring, and FDDI.

Ethernet uses a technique called *carrier sense multiple access with collision detection* (CSMA/CD)—it tries to avoid two devices trying to talk on the bus at the same time, but has rules for resolving such conflicts when they do occur. Ethernet started life at the University of Hawaii in the late 1960s and was picked up in the early 1970s by Digital, Intel, and Xerox, who developed it into the de facto standard for factory control and factory data collection markets. Classic "thick" Ethernet remains a very good standard to use in factories and other hostile environments.

Following the adoption of Ethernet as an Institute of Electrical and Electronics Engineers (IEEE) standard (802.3), it was an easy interface for different manufacturers to develop and thus became popular particularly in universities and research environments, alongside another seed element of the open systems philosophy: the Unix operating system.

With the introduction of PC networks and OSI, Ethernet has grown to be one of the most widely used standards. The most recent implementations, notably the 10BaseT twisted-pair standard, depend on signal processing in intelligent "hub" units and are quite different from the original thick coaxial design. 10BaseT uses a radial architecture that makes network management and fault isolation much easier.

The *Token Ring* standard was introduced by IBM and also became an IEEE standard (802.5). In this case, a "token" is passed round the network and a device can only talk when it has the token. Token Ring remains a relatively proprietary product, but is widely used in PC networks, particularly when they have a link with an IBM mainframe or midrange system. It does have some advantages over Ethernet for large PC networks in that it is easier to size accurately and is less prone to serious degradation under peak loads. Most of the major network software systems support both Ethernet and Token Ring.

The *FDDI* also forms part of the IEEE 802 series standards. It defines a network operating at 100 Mbps (compared with the 10 Mbps of Ethernet or the 4 or 16 Mbps of Token Ring). It is therefore most commonly used for the backbone of a network covering several floors in a building or several buildings on a site. The fact that it neither generates nor is susceptible to electrical interference can be a significant benefit in some environments.

12.1.2   LAN Software Standards

On the software side, two operating system families have dominated the PC network market: Novell's NetWare and the LAN Manager family, developed by Microsoft and IBM and now marketed by IBM. Other important PC network operating systems are Banyan VINES and an increasing number of peer-to-peer networks, such as Microsoft's Windows for Workgroups.

A *peer-to-peer* network, as in Figure 12.1, makes each user's files available to other users on the network (subject to the first user's agreement). Peripherals are also made accessible across the network, but when any of these network facilities are used, the person "lending" the facility suffers a performance degradation, which can be considerable. The *server-based* network shown in Figure 12.2, on the other hand, holds its public data and most of its shared facilities on a separate machine (the server). This requires an extra machine, but each user normally receives better performance. Either network can be implemented using a ring structure or a tree structure (described below).

From a security point of view, peer-to-peer networks are very difficult to manage, since they depend on each user granting permission to other users to share data held on his or her machine.

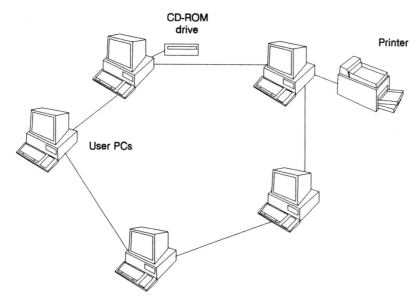

Figure 12.1 Peer-to-peer network.

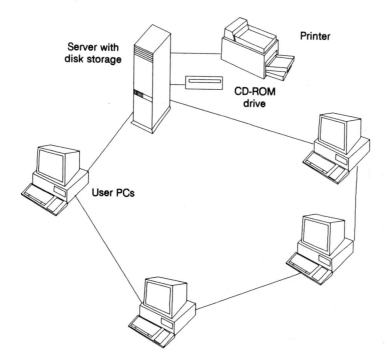

Figure 12.2 Server-based network.

12.1.3   Wide-Area Network Standards

For WANs, the hardware connection is usually made by modem (using one of the ITU V-series recommendations) or by the ISDN. Some users may have direct connections to X.25 packet-switching services using a local PAD, and two other technologies that are rapidly gaining ground are those for asynchronous transfer mode (ATM) and frame relay.

Several cities in Europe and the United States have set up metropolitan-area networks (MAN), which offer a higher capacity switched connection than is possible with the more common X.25 or ISDN service. Very-high-capacity direct links (up to 34 Mbps) are available directly from the PTT or licensed telecommunications operator.

The software is likely to conform to a manufacturer's proprietary standard, most often IBM's SNA or 3270 terminal standards, or to the common OSI standard Transmission Control Protocol/Internet Protocol (TCP/IP). TCP/IP is supported by most major manufacturers, including IBM and Digital. It offers relatively few security services.

Some WANs make extensive use of the ITU X.400 Message Handling Service standard. This standard incorporates a wide range of security services covering proof of origin and destination, message authentication and integrity checks, nonrepudiation, and encryption.

12.1.4   Network Management

Another widely used standard in networks is the SNMP and its more recent update SNMP2. These were mentioned in Chapter 10, and should be supported by most current network hardware and software. Network management is critical to any security-conscious installation. Although SNMP is somewhat restricted in scope, the alternative Common Management Information Protocol (CMIP), which is promoted by some OSI groups, is much more complex and supported by fewer devices. At present, SNMP and SNMP2 are likely to be more useful.

12.2   INTERNAL NETWORKS

12.2.1   Network Architecture

A very high proportion of all computer networks are used within a single company and within a single building or factory complex. This kind of network is classed as a LAN, as opposed to a WAN, which uses cables or other communications links provided by other organizations. A LAN consists entirely of direct connections, and so it can operate at quite a high speed (up to 100 Mbps, although 1 Mbps is common). Modern LAN hardware is intelligent (firmware-controlled) and uses backbone and star architectures (see Figure 12.3), which make it much easier to detect and isolate errors and to reconfigure the network when people or requirements change.

Overall network reliability is improved by using this type of architecture. There are also standards to which these modern *structured-cabling* installations should conform: the Category 5 standard, used for structured data and telephone cabling, is the most popular.

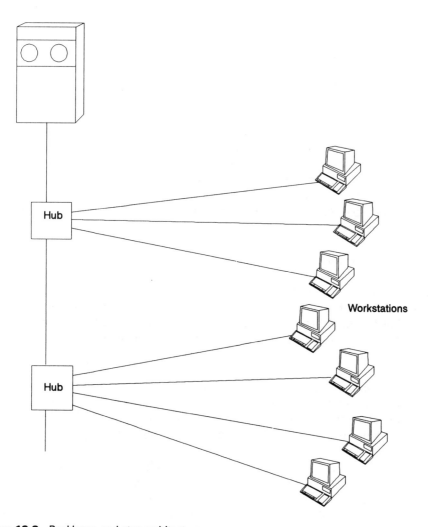

Figure 12.3 Backbone and star architecture.

12.2.2 Applications

A company's internal LANs are used for a very wide range of applications, some of which are very specialized and specific to the company or sector. Common generic applications include:

Word Processing. Each user usually keeps his or her own document directories, or a secretarial group may hold directories for each member of staff. Great care needs to be taken with documents that may need to be accessed by several members of staff, not least because different setups on their respective systems may cause the documents to look completely different on printout or display. A small number of people should control common document appearance standards, and office procedures should normally dictate who can originate, check and revise, read, copy, or print a document. These procedures must be reflected in users' access rights.

Spreadsheets. The same general principles apply with spreadsheets, except when users are restricted to running certain macros or spreadsheet programs (most macros are interpreted rather than compiled, so users can find ways of exiting into the main spreadsheet software). It seems to be more common for managers to maintain their own spreadsheets than their own documents, despite the dangers this can bring if the full consequences of a command are not understood. Companies dependent on spreadsheets should have a specific policy for this issue.

Databases. It is much more common for the database and access to it to be controlled and available to most staff only through specially written compiled programs. We deal with the application of databases in the next chapter, but it is worth mentioning here that the network architecture must be set up in such a way that the integrity of any protection is maintained; there is no point in restricting access to a *program* if the *data* are available to a standard package on the user's local PC.

Standard Application Packages. Standard application packages (such as accounting packages and production control suites) are dealt with in more detail in the next chapter, but the same warning about the need to restrict access to or otherwise protect data as well as programs applies in this case also.

Electronic Mail and Other Messaging Systems. Those with experience of electronic mail (email) systems restricted to one company or internal network will know that security is not the main issue here. Discipline in use of email consists mostly of making correct choices of addressee (not sending every message to everybody) and in deciding when to attach a file or incorporate it in a message, or simply to give a pointer to it. Some care needs to be taken over the choice of names in an email system, particularly in a larger organization where the chief executive may share a surname and initials with a short-term contract worker (British Telecom suffered an embarrassing security breach in this way: the contract worker turned out to be a reporter on a local newspaper). The common habit of using names like "JackS" for Jack Smith increases the danger of this type of problem. Even on internal networks, some outsiders (subcontractors, suppliers, or even customers) may have mailboxes. This requires even more care with global messaging, and an intelligent choice of group names is essential.

12.3 FIELD SALES NETWORKS

12.3.1 Requirement

Staff who spend much of their time outside the office, such as field sales forces, service and installation engineers, and inspectors, need access to the company computer system. Most of the requirements are similar for all these groups, and we can treat them together under this heading.

Field sales networks are used for maintaining sales leads and customer records; planning visits; and recording staff time, movements, and expenses. They are often the main or only communication the salesperson has with the head office.

Many field sales forces nowadays have portable PCs that have either a built-in modem or a modem they can use at home in the evening. The PCs themselves are vulnerable to theft and damage, and it is very difficult for the company to control how many other applications the salesperson finds for the PC and modem.

12.3.2 Controls

Key controls are therefore on the data and on the process by which salespeople log onto the internal network to exchange their messages. It is also wise to have some form of bootstrap control on the PC so that a casual thief or person gaining access to the PC will be deterred from investigating the data and programs.

Important customer data should normally be encoded or encrypted so that it is not readily accessible to someone gaining access to the file or monitoring the transmission. The decoding or decryption should be carried out within the application program, where it is much easier to implement security checks. If the data are protected, there is then less need to encrypt the transmission. It is very undesirable to have relatively expensive encryption equipment at each salesperson's home.

If the exchange of data is always to be carried out from home, then a dial-back

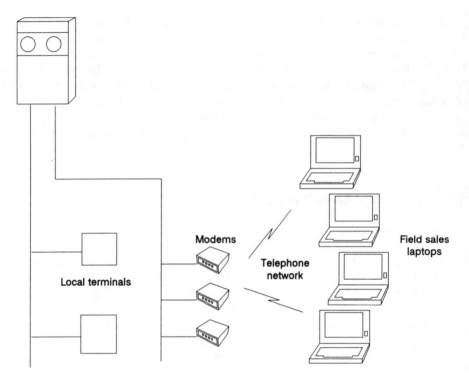

Figure 12.4   Typical field sales network.

system will be the most effective method of controlling the log-on procedure. When calls may be made from a hotel or from customer premises, however, this is not viable, and in this case a token-based log-on procedure (see Section 9.5) would probably be recommended. It would be important for the salesperson to keep the token separate from the PC, even when a PIN is used to initialize the token. See Figure 12.4 for a standard field sales network configuration.

Having a field sales network automatically makes a company network more open, so logical controls become more important than physical ones. User applications as well as data must be protected from other users, and serious consideration should be given to upgrading the operating system or installing any optional security modules that may be available.

12.3.3   Radio Networks

In addition to the modem option described above, mobile staff in some sectors use radio networks to communicate with the head office. This is most often required when there is a real-time element or frequent requirement for access to a database, as in police forces and utility companies. The radio network may be a private mobile radio (PMR) system using dedicated frequencies, or a cellular packet radio system (such as one of the Mobitex standard networks). In these cases, the radio system and linked database are usually set up as a separate system with defined links into the main computer system (see Figure 12.5). The protocols for transmitting digital data over a radio system normally involve some hashing of data for forward error correction purposes; although not as secure as encryption, this will prevent casual eavesdropping. Since the transmissions are by their nature randomly timed, the risk of successful monitoring is generally quite small. The data stream can be additionally encrypted if required.

Any medium that has no permanent connection between terminal and network relies on some form of identification check when the connection is established. In radio modems a personality chip is normally used to establish a first level of identification, which may be complemented by a single or two-level password entry.

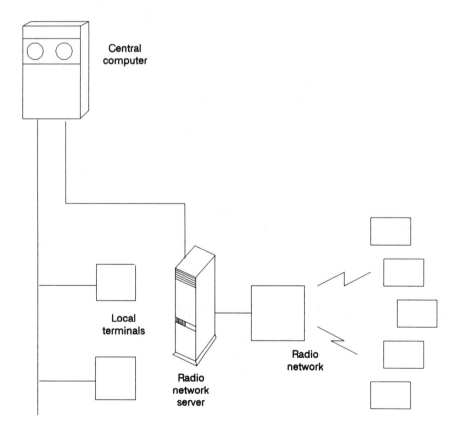

Figure 12.5 Mobile radio network.

12.4   ENGINEERING AND DESIGN

Engineering and design networks are at the other end of the scale from the field sales network. Here the requirement is for top-end PCs or powerful workstations used by specialists in, typically, a fairly restricted area within the company. Physical controls are therefore very appropriate, although they should be matched by access rights controls.

Unlike the standard commercial network, where users run programs held on the server or mainframe system, the design suite is likely to be held and run on the workstation or local PC. Depending on the number of designers involved, the data may be held locally, centrally, or divided between the two.

Unlike the standard commercial systems, many CAD programs are regarded as valuable in their own right and may not be copied even for security purposes (system backups would normally be excluded from this). Dongles or other forms of protection are often used.

CAD files have traditionally been specific to the CAD system being used, but increasing numbers of systems can now import data from other popular programs, and there are two or three standards competing to be used as a common interface between CAD standards, culminating in an ambitious project called STEP, which seeks to be able to portray all aspect of a product, from the materials and standards used to design, manufacturing process, and conformance testing. Like the equally broad computer-aided acquisition and logistic support (CALS) standard promoted by the U.S. Department of Defense, STEP information is most often exchanged on magnetic tape. But there is a demand for a secure network system to exchange these large files, and through the use of ISDN and other higher speed switched networks, the technology exists to support such a network. We therefore believe that network planners should today already be taking this type of large file exchange into account. Like most EDI standards, STEP handles data in clear form. When STEP or other design data are being transmitted across public networks, encryption should always be considered. Any network regularly handling this type of exchange should use both encryption and authentication.

12.5   INTERCOMPANY DATA TRANSFERS

Contrary to the impression that might be gained from reading the nontechnical (and even some of the technical) press, the vast majority of data transfers between companies today take place on magnetic tape or, increasingly, floppy disk.

Most of the data are unencrypted, and the tapes are sent by couriers often unknown to the sender or receiver. The data are used quite trustingly by the receiving company; often they will be read into a file where some basic checks are carried out only on the format and header before full processing starts. In practice, this apparently very insecure system is responsible for remarkably few of the security breaches that do occur. The worst (actually quite common) situation is that the data are found to be unreadable or in the wrong format for the receiving company, resulting in delay and inconvenience rather than any loss of data or integrity.

Data communications are supposed to be able to eliminate these delays and errors. Again, practice is a little different. Widely varying modem standards, telephone exchange systems, and file transfer protocols all conspire to make one-to-one file transfers between consenting adults remarkably difficult. Even when these differences are overcome, the data may be no more reliable than data on a magnetic tape. Such file transfers are therefore most often used on an ad hoc basis, when there is a need for urgency or where postal or courier services are unreliable.

More frequent file transfers between two separate networks now often take place by email and EDI systems run over public or private networks. It is much easier to send an email message, with a file attached to it, to a customer or supplier through a server to which both companies have access, than to send that file or message directly to the other company's system.

12.6   PRIVATE VANs

12.6.1   Characteristics

It was largely to meet this type of requirement (for regular file transfers between separate systems) that the first VANs were set up. The VAN took away the problems of making the connection, and the two users only had to agree on a format. Many users of IBM's Information Network or AT&T's EasyLink still operate in this way. In the early days, the VAN operator required that all users use certain approved modems (to eliminate one variable). Nowadays the differences between modems on the common V.22, V.22bis, V.32, and V.32bis standards are minimal and can be accommodated by all good software. (The same is not yet true at the higher speed 28.8-kbps standards.)

What is still likely to be common, however, is that the software package is supplied by the VAN operator. In this way the operator ensures reliable connection and diagnostics. The difference between a good network software package and a standard communications package is as much in the logging and diagnostic functions it provides as in any user features.

The VAN itself is also likely to add security and reliability features, such as logging and incident trapping, error recovery, and redundant paths, as well as the more obvious features sold by the network, such as local call rate access, directories, and added-value functions such as store-and-forward messaging.

12.6.2 EDI

One of the most valuable functions a VAN can provide is EDI. Although EDI does not depend on the use of a VAN (only on the use of an agreed-on set of message structures and data fields), nearly all the EDI schemes of any significance today run through a VAN. With the increasing availability of ISDN, this could change, but this would imply a change in the way most EDI schemes operate.

EDI is the use of direct links between computers to transfer data that would otherwise usually be sent in printed form (see Figure 12.6). The data are nor-

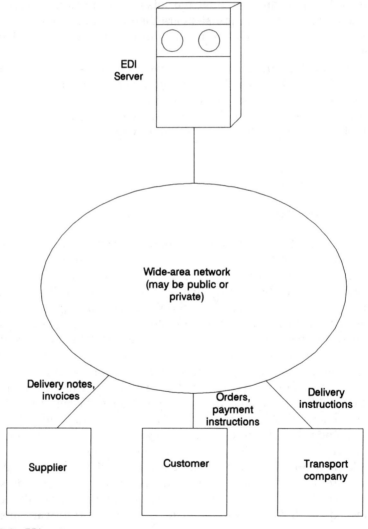

Figure 12.6 EDI systems.

mally structured using standard message formats and data elements, and one of the key elements of an EDI scheme is the agreed-on set of standards.

In Europe, most new EDI schemes use the EDIFACT standards (or others such as EANCOM which are based on EDIFACT), although there are still many schemes, such as the U.K. Tradacoms scheme, which are based on older standards. In North America, the ANSI X.12 set is the norm. There is much common ground between EDIFACT and X.12, and translation between the two is relatively straightforward. EDI links applications in the two systems; it therefore tends to sit on top of an OSI system rather than form part of it. The syntax of both EDIFACT and X.12 are different from that of the preferred OSI messaging standard X.400, although the X.435 extension has been defined to allow the EDI messages to be contained within an X.400 message structure.

EDIFACT and other EDI standards define messages using printable characters only, and many EDI operators discourage or disallow encrypted fields. Some security services are defined in the EDIFACT and X.12 standards, and users of the X.435 extensions can also take advantage of the X.400 security services, including a MAC for the message content. For more sensitive data, a coding scheme (rather than encryption) can be agreed on between sender and recipient. For example, product codes can replace descriptions and location codes can replace addresses.

The most common application of EDI is in paperless trading: the transmission of purchase orders, invoices, delivery notes, and customs documents between a purchaser, a vendor, and their transport companies and customs authorities. EDI is also, however, used in banking and financial applications, as we will see in Chapter 14. These applications are very much more demanding of security and so normally make use of proprietary standards.

12.7 Public Networks

A public network, in principle, is one to which everyone has access. In practice, the distinction between public and private networks is becoming blurred:

- Public networks used to be run by national PTTs. Although state-owned PTTs do still exist in many European countries (despite strong pressure from the European Commission), their data networks were among the first of their activities to be opened to competition, and most PTTs now run their data networks as fully commercial operations.
- Many of the large private networks, such as CompuServe or America Online, have no joining restrictions and operate very much as public resources.
- The public networks were originally "no frills" networks that offered only a basic voice-grade connection without any added-value services. Advances in technology and the need to meet competition have changed this

dramatically, so that most of the former PTT networks can now offer a full range of services and line qualities to equal their private competitors.

The main features of the public network today are that there are few or no joining restrictions and no control over the software used. Security at the log-on stage is therefore minimal; anyone who makes information available on such a network must assume that anyone can gain access to it—this is, of course, usually the desired effect. When a company makes an application available through a gateway on such a network, the application must take full responsibility for its own security.

12.8   WIRELESS AND SATELLITE

One of the most important trends in networks in the mid-1990s is the increasing use of wireless technology. Ten years before, it was thought that most international communications would take place by satellite, since the cost of top-quality copper cables was prohibitive for long distances. Fiber optics have changed that view considerably, and nearly all national and international telephone traffic now passes through this medium. There are, however, many situations in which wireless communications are used.

Point-to-Point Communications

In the United States more than elsewhere, there is considerable use of microwave links for high-speed communications between two points. VHF and UHF links can also be used as well as optical communications for short distances. Each of these has its own characteristics with regard to maximum distance and the extent to which a clear line of sight is required. Generally, lower frequencies will pass around larger obstacles, but even the higher frequencies can operate without true line of sight because of multipath effects, which cause the signal to be reflected from various surfaces along the way. Even broadcast radio and television signals are reflected from the upper levels of the atmosphere. The network designer has to be aware of the ways in which these reflections can change: leaves, for example, reflect much more when wet. UHF signals are more affected by passing airplanes than lower frequencies. Put together, these effects can alter the characteristics of a radio transmission system, particularly a focused point-to-point system, quite considerably.

Narrow-beam transmissions are rarely so narrow that they cannot be monitored. Companies using such links should assume that others can monitor them, and when the data are sensitive, they should take suitable steps, including encryption where necessary. Because of the volume of other traffic, monitoring would be unlikely where a microwave link forms part of a public network used for one leg of the transmission. Systems using direct point-to-point radio or microwave links to link two buildings, for example, must be aware of these effects and where necessary take greater precautions.

Private Mobile Radio (PMR)

Taxis, emergency service vehicles, and other commercial vehicles use PMR frequencies, usually in the VHF or UHF bands, to transmit data as well as voice. These bands are very congested and a high level of interference is often experienced. The design of the coding schemes must take these factors into account, and a wide range of radio data hardware is now available with these facilities built in. Normal users can assume that the resulting communication is fairly secure as well as reliable, although the police and other agencies with particularly high security requirements will take additional precautions.

Low-Power Radio

In the past few years, a complete range of low-power radio communications devices have been exempted from the need to obtain licensing approval for each installation. The devices normally have to be type approved under national legislation (unfortunately the frequencies available are different in each country), and are likely to be limited to a transmit power of around 1W.

They do, however, open up a range of possibilities for short-range communication in environments where a degree of mobility is necessary or where cabling is impractical: for waiters in restaurants, for example, or in retail shops where the owners are unwilling to allow marble floors or mahogany paneling to be disturbed for new point-of-sale systems. Many office computer networks have been designed using low-power radio simply because of the flexibility it offers.

Again, there is a need to take into account the way the transmission characteristics may change with time, as the shop fills with people, for example, or as desks are moved around in an office. Such networks should therefore be operated well within their technical limits to avoid problems when short-term changes reduce the quality of the link.

Cellular Radio

The explosive growth of mobile telephones may have obscured the fact that cellular techniques can be valuable for data also. Analog mobile telephones can be used with modems, probably up to 2,400 bps with success.

With the move to digital cellular systems, the likelihood of successful data transmission at much higher speeds opens up. The GSM standard is now well established throughout Europe, with networks operating in most countries. It has also been introduced to many other parts of the world and seems to be the only standard with a chance of worldwide acceptance in this field. Unlike analog cellular radio, which is notoriously insecure, GSM includes quite a high standard of security through the use of encryption and a "personalization chip," which contains data used in identifying the user at call setup. Although it would be difficult to prove that an eavesdropper could not extract a sequence of keys, the chances or resources required to reconstruct a complete message would seem to be beyond practicality.

Another important standard for data transmission using cellular technology is the Mobitex standard, developed by Ericsson for the Swedish PTT. This has been licensed to a number of operators in different countries, and the standard is now controlled by Mobitex Operators' Association. Mobitex also assures a high level of data integrity, and at present the number of receiver manufacturers is small enough for a reasonable level of control to be enforced. Users with sensitive applications would nevertheless need to take precautions against eavesdropping.

Satellite

Satellite communication is very topical at present, with in excess of 100 commercial communications satellites now in use, and at least a further 10 being launched every year. The most important satellites are those in geostationary orbit, 35,787 km above the equator. They appear to be fixed in the sky, and so an aerial can be fixed towards them. Satellites have different power levels, with the newest units being much more powerful than their older cousins.

The signal-to-noise level at the satellite receiver depends on the satellite's transmitted power, the size of the dish, and the quality of the low-noise block. The other factors are mainly the losses in the earth's atmosphere, particularly when it is raining. These are quite predictable, and so the performance of a satellite system, once set up, is attractively consistent.

The bandwidth available on a satellite transponder is relatively high— communications satellites usually make available between 36 and 72 MHz on a 6- or 14-GHz carrier. It is usually best to make use of some spare bandwidth to provide redundancy in the data for maximum reliability, but users who have experienced radio data transmission will in general be very pleasantly surprised by the quality of satellite reception.

The accuracy with which the receiving aerial can be focused on the satellite depends on its size: larger dishes can be focused much more accurately. Small dishes, even when they receive adequate power, may suffer from crosstalk from adjacent satellites. Only the very lowest data rate signals can be picked up without a narrow-beam antenna.

Signals broadcast from a geostationary satellite can be received by anyone within the satellite's footprint; even outside the central area, a larger aerial will often pick up a usable signal. Most of the encoding techniques (usually a form of phase-shift keying) are well known. Users of satellite systems will therefore have to build suitable encryption or other protection into their protocols.

Many satellite television services make use of either the Videocrypt or Eurocrypt standards, both of which use a smart card in the set-top receiver to decode scrambled picture signals. In both of these cases, the audio signals remain unencrypted, but the "spare" signal lines usually used for teletext, known as the *vertical blanking interval* (VBI), are used for the encryption data. This is important for data transmissions, since many television signals carry data either in the audio subcarriers or in the VBI. The analog television encryption standards are incompatible with the use of the VBI for data. The European Broadcasting Union (EBU) standard for VBI data transmission includes, however, a conditional access mechanism based on a simplified DES scheme.

Very-Small-Aperture Terminal

Of particular interest to data communications network designers are two-way *very-small-aperture terminal* (VSAT) networks. Here the remote terminal (the VSAT), which has a small antenna (typically 90 to 150 cm), can only communicate with a hub, which has a much larger antenna. Communications between VSATs pass through the hub, as shown in Figure 12.7.

The same arithmetic of transmitted power, path loss, and receiver antenna

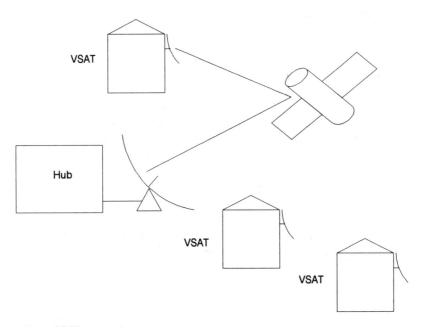

Figure 12.7 VSAT network.

size has to be performed as for the receive-only station, but in this case in both directions. Again, however, performance of the system once set up is usually quite predictable and consistent.

Low-Earth-Orbit Satellites

As geostationary orbits become very crowded, there is renewed interest in low-earth-orbit satellites (LEO), many of which have been launched over the years, but few are used commercially. LEOs can be launched much more cheaply and are likely to be much smaller and cheaper than geostationary satellites. The disadvantage is, however, that they move across the sky and may only be in sight for a few minutes each orbit. The antennas used must track the satellite across the sky, which is not a difficult job for today's electronics, but requires moving parts.

A LEO network is likely to consist of several satellites. In some designs the satellites communicate with each other in order to maximize the possible paths. LEOs are able to use quite low frequencies in comparison with geostationary satellites and are therefore more prone to interference. This could be particularly serious in the case of the signals used to control the satellite, and so relatively sophisticated code-division multiplexing or other spread spectrum techniques are invariably used.

12.9 SUPERHIGHWAYS

There are undoubtedly sectors where the very wide availability of all kinds of data creates the impression of a revolution in technology. In other sectors, the revolution relates more to the speed with which data can be transferred and used than to the availability of the data. With the exception of large graphics files and full-speed video, today's data transmission speeds meet or exceed the speed at which companies are able to make decisions or products. Even in the very-high-speed and technical world of interest rate and foreign exchange arbitrage, it makes little difference whether data are available in millionths of a second or tenths of millionths. Ever higher speeds will continue to become available, but this is unlikely to affect the efficiency of the majority of companies, still less a nation.

The most discussed network with a claim to superhighway status is the Internet. The special characteristic of the Internet is the lack of any controlling organization. A few rules and protocols were set down at the outset, but now the network depends only on the participation of its members. The fact that so many academic networks are linked into it makes it very large in terms not only of terminals and registered users, but also of the quantity of data, particularly publications and abstracts, available through it. Its academic roots also ensure that there is a substantial base of anarchically inclined and technically very skilled users who will be able to beat any half-hearted attempt to enforce security.

Some organizations offering Internet access to commercial users take the security issue seriously, but those taking advantage of these services need to examine carefully the protection offered in relation to their network. Organizations with sensitive data or applications should consider whether the requirement for Internet access is in fact genuine. If it is, then some form of fire wall should be built: any software used for accessing the Internet should be carefully controlled and its suitability checked. An easy route is to restrict access to the Internet to PCs not linked to the network.

Other, more commercially oriented public networks (such as CompuServe, America Online, Lotus One Source, and Microsoft World) offer almost as wide a range of connectivity and data, while the user is protected by contracts and the network provider's commercial interests as well as by the user's own software and access rights mechanisms.

Part III: Applications

Commercial **13**

13.1 PROCESSES AND DATA

13.1.1 Transaction Processing

In the next few chapters, we will be considering those cases in which the nature of the risk and the measures that should be taken to counter those risks are dependent on the nature and content of the data. In this chapter, we consider standard commercial applications, applicable to a wide range of companies in manufacturing, industrial, and professional services. These systems are increasingly *transaction-oriented*; that is, they process data or requests as they come in, rather than batching them up for processing at the end of the day or once an hour. Transaction processing places more stringent demands on security, since the results are being acted on almost immediately, while the overhead involved in checking the validity and correctness of every question and answer is much greater than it is with batch processing. Administrators concerned with setting access rights for a network must bear in mind the type of processing involved as well as the nature of the application.

A full formal model of the security requirements of a commercial transaction processing system has been developed by Clark and Wilson [1]. Reference to a model such as this is required for products seeking ITSEC approval at levels E4 and above, and would be recommended for any organization with specific high-security requirements even in the limited area of its operations.

13.1.2 Databases

As we said in Chapter 1, it is usually the data that have value and require protection rather than any program or process in itself. We will mention a few exceptions at the end of this chapter. Most computer systems are structured so that programs and data are kept apart. Nowadays differences between one installation and another are dealt with by parameters: a *configuration file*, which is set up and used by the program, determines what computer and network are being used, what types of terminal or printer, and other user preferences. In this way, the program itself never changes; it can be used by many users on a network, and it can be checked by virus detectors to ensure that it has not changed.

Data, on the other hand, are continuously changing, and the way they are shared between users on a system is very important. Multiuser systems and networks have to have a system for sharing data, and this is done by *file and record locking*. A program that can make changes to a data item should lock the record containing that item between the point where it reads it and the point where it writes it back. When a database system cannot handle record locking, the whole file must be locked. This is a much less efficient way of working and shows up when users repeatedly see messages like "Unable to access file."

Modern systems are generally structured as *databases*, even if purists would often dispute the definition. A database is a collection of files and links between these files, usually in the form of common (linking) fields, key fields, and key files.

Relational databases (again a pragmatic rather than a pure definition) allow various views or dimensions of the data in the files. In a company database, for example, it might be possible to view the hours worked by employees who are under 20 years old, even though the age and hours are kept in different files. Different parts of a database may have different access rights, and the software used must support these varying rights. Databases of this type are often used for employee records, customer files, sales leads, and stock records.

As a general rule, database access should be restricted to compiled applications: the payroll data to the payroll program and the time and attendance program, and employee dates of birth to the employee records program. When special reports are required (such as the hours worked by younger employees), a structured query language or report generator program should be used. Companies following this approach should, however, bear in mind that when standard database formats and languages are used, there is usually a basic package that will allow all operations on a database; this is certainly true of all the PC database packages. An operating system with a good server directory protection mechanism (see the comments on operating systems in Chapters 10 and 11) will prevent unauthorized users from accessing specific files. In other cases, any sensitive data should be held on the file in encrypted or encoded form, to be decrypted or decoded only by the dedicated program.

When the database is set up, it is important for the designer to know who will have access to different items of data. Items with different access rights limitations should be held in separate files, even if this means a very large number of files.

13.2 PERSONNEL RECORDS AND PAYROLL

In many companies, personnel records are regarded as highly sensitive. Certain aspects may be governed by data protection legislation in some countries.

Sensitive areas include:

- Salary structures and bonuses: Attitudes to these vary greatly according to cultural factors. While in the United States, for example, a senior manager's high salary may be a matter of pride, in Europe it is more likely to be an embarrassment. Information on directors' salaries is normally given in annual reports of public companies, but this does not always allow the individual salaries to be determined. Bonuses are even more complex. Again, some top salespeople boast of their bonuses while others try to avoid causing resentment. As a general rule, it has to be the employee's choice whether his or her salary or bonus is made known, rather than that of the computer network. Company policy should determine how far line managers should know the salaries and bonuses of their employees.
- Annual reviews and other comments: These can be even more sensitive, not least because some comments could cause the company legal or employee relations problems. They should be available on a strict need-to-know basis in accordance with company policy.
- Employees' previous history and reports of social problems, including criminal records: These will normally arrive in paper form, and unless there is a good reason for entering them on a computer system, they should probably stay on paper. With the increasing trend towards wide-area email, companies may be tempted to ask for and send references by email. Such mail should be treated with the utmost caution and should probably only be used in a privacy-enhanced mail system.
- Age and personal position (including tax position): This is not usually highly sensitive information, but should nonetheless be reserved to the personnel department and direct line managers.

Fortunately, it is rare that employee records pass outside the company. Payroll details, however, are quite often handled by a separate company as a bureau operation, which can involve significant data exchanges. Since this is normally a batch operation (even where time and attendance data are required, it is taken to the end of the previous day), a secure file transfer protocol can readily be arranged. The form of connection between the bureau and the company should be considered carefully. Leased lines or connections via a private network are relatively safe, but where dial-up, public packet switching, or ISDN lines are used, the log-on process must provide good protection: one of the mutual authentication schemes discussed in Chapter 10 should be used.

Most payroll bureaus will be very aware of security. In the 1970s there were one or two successful computer system sabotages involving disaffected employees and payroll files, but this kind of threat has now dropped to insignificance, since security has been increased and more companies handle their payroll inhouse using a dedicated package.

13.3   CUSTOMER LISTS AND SALES LEADS

13.3.1   Background

Sales staff are often on low salaries and earn most of their income from commission, leading them to have low loyalty to the company and a very competitive attitude toward one another. They also often work under time pressure and have a series of goals that can directly affect their bonus. Computer and security staff are usually aware of these circumstances in principle, but often underestimate their importance for the system. Because the sales staff are often out of the office, their networks tend to have the lowest security in the whole company system, but they nevertheless do have a high need for sensitive data. Even call plans can be remarkably sensitive. In one instance in Germany, a competitor obtained a copy of a pharmaceutical salesman's call plan for the week. By arriving a few hours earlier in every case, he was able to secure the lion's share of the available business in that area that week.

As we discussed in Chapter 4, a realistic assessment must be made of the various risks involved and the level of threat arising from each risk. A balance must be drawn between the risk and the cost, in time and equipment, of the necessary countermeasures.

13.3.2   Customer List Copying

Customer lists are probably the most vulnerable item in any computer system. A very high proportion of salespeople leaving a company will, if they have any form of computer access, take at least part of the customer list with them, even if only by printing it out. It is impossible to prevent this completely. Salespeople's capital consists to a large extent of their contacts. They will probably have business cards, notebooks, and other sources for much of the information anyway. They need access to the files for their normal work, and, until the moment they resign, must be assumed to need it on valid business. Nor should the system make them jump through too many hoops in order to gain access to the files: many salespeople are impatient with computer systems and want quick answers. An atmosphere of distrust creates distrusting employees and positively encourages staff to find their own ways round the protection. The best solution is therefore to accept that the salespeople will have access to their current contacts. The computer system can, however, reasonably prevent anyone from copying the whole file to another disk or to another computer. When field sales staff have their own computers, this is a little more difficult but well within the capabilities of a good access control system or package.

Many employees leaving have plenty of time to copy the data in small quantities, but it may be a good idea for the system to track the number of records accessed over a week or a month, say. This may be a useful management indicator as well as a source of exception reports. When a company employs several salespeople, the system should also prevent one salesperson from accessing details of another's customers. This is not only to prevent data theft, but also to maintain fair and normal competition between salespeople.

13.3.3 Sales and Marketing Systems

PC-based sales and marketing systems (see Figure 13.1) can have a dramatic effect on the efficiency of a field sales force. Laptop computer-based field sales

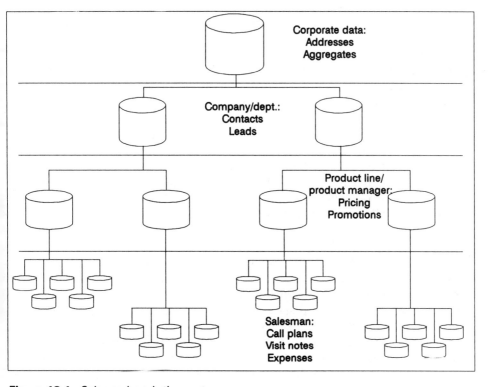

Figure 13.1 Sales and marketing system.

networks were discussed in the previous chapter. They raise a number of security issues that must be addressed.

Salespeople make their calls during the day following a call plan, take orders, and make notes on the visits. In some sectors, it is now acceptable for salespeople to use a PC during the meeting; in others, the orders and notes must be written up after the meeting, in the car. At the end of the day, any further notes (expenses, leave requests, or other messages) are added, and the computer is plugged into a modem, telephone point, or ISDN adapter. The software system automatically connects to the company computer center or private network, uploads the day's changes, and downloads any changes, messages, or call plans made by sales administration.

The first specter this raises is that portable PCs can be very easily stolen. They are frequently stolen from salespeople's cars, from their briefcases while they are making visits, and from their homes. A good access control package, with encryption or coding of critical fields, will ensure that the data are safe, although the inconvenience and loss of time involved can be considerable.

The sales and marketing package must be very selective. It should record changes at the record level, and records must include the salespeople's references, so that each salesperson only receives his or her own information.

Most such networks today operate on the public switched telephone network, even where transmissions only take place from salespeople's homes. When transmissions from hotels or customer premises are a possibility, a dial-up option must at least exist. Telephone numbers for these dial-up nodes should be built into the software and ideally should not be accessible even to the employee. In practice, modern call logging makes this almost impossible.

The password system can, however, be made very secure, preferably through the use of a separate token or a variable field in the log-on dialogue. Transmissions should be signed and authenticated, with a new key or an element of the signature block being transmitted within every exchange. Dependence on manual input of a single-level ID and password should be avoided in this type of network whenever possible. When it is unavoidable, very tight password discipline must be maintained.

13.3.4 Data Content

The vulnerability of the data can also be greatly affected by the way they are held. Customer categories can often be held in coded form, with the code lists held in a separate file. The system protection can probably ensure that the code list can only be accessed by the display or input programs. Even address lists can be held in very compact form so that they can only be readily decoded by a program. In most countries there is available a comprehensive postal address register, including postal codes (zip codes). Frequently used words (including *street*, *road*, and *west*) can be held in a list that is accessed by a limited number of programs.

Details of leads and orders should again be held in files separate from customer lists. Customer codes can form the link, perhaps with an algorithm being used to convert from the one to the other. It must be possible to link orders with customer names in the accounting system, where a similar link must exist.

13.3.5 Contact Packages

Contact packages are another aid to efficiency, which can be used both in internal and field sales environments. They are usually less comprehensive in their scope than the full sales and marketing support systems, and they are less tailored to the specific company environment. But they handle full customer lists, keep track of all contacts and discussions, and can provide call-planning functions.

The fact that they are relatively inexpensive and widely available increases the risk in using them to store sensitive company data. Even where codes are used, they must be stored in a format that can be interpreted by the package. It is not normally possible to encrypt files (since the package accesses individual records). A specific network version of the package may offer slightly more security, but in general the range of security or data communications options is not wide.

13.4 ACCOUNTING

Most organizations are aware of the sensitivity of their accounting information. Cultural factors greatly affect companies' view of the need for secrecy of accounting data. Although it is rare that unauthorized disclosure of a single item of a company's accounts would do the company significant commercial harm, many organizations take the view that accounting information, taken out of context, could damage the company or its relationships with its customers.

Although many parts of the company play a role in building up the accounts data (see Figure 13.2), the accumulation and presentation of the data in a set of accounts are strictly the preserve of the accounts department. In a large accounts department, the sales and purchase ledger groups may not even be allowed to access each other's data.

In general, individual cost items are rarely regarded as confidential. Costings of complete equipment or items for sale may be more sensitive, since these allow determination of the company's margins. When the selling prices of equipment or services are subject to negotiation, they are much more likely to be treated as confidential.

Unauthorized alteration of data in accounting systems represents the largest threat. Not only does this offer opportunities for fraud, but it can actually be life-threatening for the company. In addition to all the access control and data

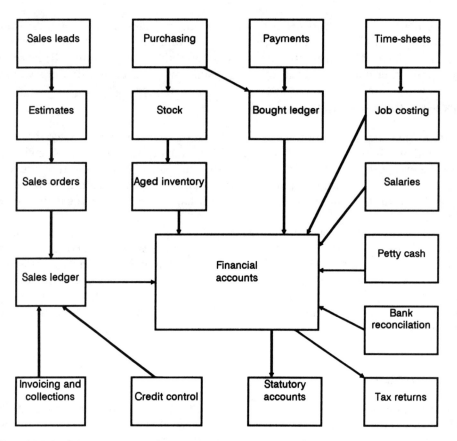

Figure 13.2 Company accounting systems.

management functions discussed so far in this book, two specific approaches can be taken to provide more security:

- Operations such as opening new ledger accounts (a popular way of preparing the ground for fraud) should be logged, and the log should be inspected as a standard operational procedure. The first operation on a new account should also be flagged and, if necessary, form part of an exception report.
- Financial ratios, which are widely acknowledged to be among the most useful financial tools in the management armory, should be extended so that they cover every line of the accounts. For companies whose business is very stable, time-series ratios can also be very telling (the ratio of this month to the average of the last 12). Unusual operations will nearly always show up through a significant change in some ratio.

The more completely the company accounts are integrated with its operations, the more difficult it becomes to perpetrate fraud. If an invoice cannot be raised without a dispatch note, for example, it becomes difficult to raise bogus invoices. Quality standards such as ISO 9000 are often felt by companies to increase overhead unduly, but they have the side-effect of increasing security significantly.

Data communications requirements in the accounts department normally relate to EDI or to electronic banking. The department prepares lists of invoices to be paid, which are then submitted either to the EDI system via a file translation program or to the electronic clearing system via special software. The second requirement is dealt with in more detail in the next chapter.

13.5 PRODUCTION CONTROL

Production control systems in a manufacturing environment can be very complex, and they are often so closely integrated into the company's operations that a failure or malfunction can cause a major loss of production. The biggest danger here lies in the complexity of the system itself and the way the operators interact with it. With well-established software and good operator training, the main remaining factor is the stability of the network itself. Here it is critical to ensure that there is enough redundancy that no single component failure can bring down the network.

Much production control software is developed inhouse. Here the problems of secure software development come to the fore, and the risks arising from poor control can be considerable. One of the main problems is often the analysts' and programmers' lack of knowledge of the application and of the working environment. In this instance, formal software development methodologies and structured programming techniques are almost essential.

Software written for this type of environment should be highly modular, since changes can be frequent, particularly in the early stages of implementation. It is then an advantage if the data can be similarly subdivided. Each module can then carry out consistency checks on its input, thereby preventing errors from propagating throughout the system. Many production control systems are now based on PC networks (traditionally they were always based on minicomputers), and the distributed architecture can be used to enhance system reliability still further. Figure 13.3 shows a typical production control system making full use of distributed processing and task separation.

Unauthorized access is only normally a problem when an untrained operator tries to carry out an operation and makes an error. Again, good software will detect the error and will prevent the operation, or at least draw attention to it and allow recovery. Software that is not so good may not only allow the error to persist and find its way into production, possibly affecting other systems on

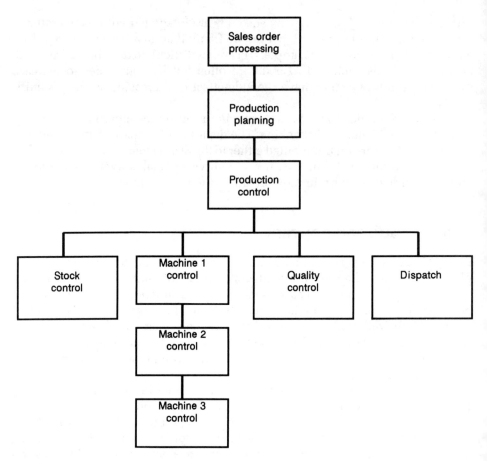

Figure 13.3   Distributed operational system.

tho way, but the program may even come to a halt or cause a serious error at a later stage.

It is rare that production secrets (such as formulations or test results on new products) are directly accessible from the production control system. Sometimes, however, the same hardware or network may be involved, and in these cases the different nature of the data must be recognized.

Production control systems have been the subject of sabotage attempts, but these are usually fairly unsubtle and physical, involving large hammers, wire cutters, or fire.

13.6   INTELLECTUAL PROPERTY

Company networks are now widely used to store confidential data, including designs and formulations, programs, and other copyright material. Many people will need access to this material in the course of their work. So although the usual access control criteria should be applied (generally on a need-to-know basis), they will be insufficient to prevent knowledgeable insiders from printing, or perhaps copying, the material. Users at dial-up ports or ports accessible from an external network can and should be prevented from gaining any access to this type of material, however.

In order to impose any form of control on the intellectual property on the network, the first requirement is to know that it is there. It is perhaps surprising how few companies have a register of confidential, copyrightable, or patentable material stored on the network, and yet this is an essential prerequisite and can be a valuable exercise in its own right, drawing management attention to the nature and value of the information held within the company. It is true that the register itself then becomes a document of particular interest to the industrial spy or casual intruder on the network, but a single file can always be comprehensively protected. It should then be possible to ascertain, from the normal access rights in use, who has access to the information. This may reveal a need to change the access rights or to put special controls on these files, including encryption if necessary.

If the operating system will allow it, the next stage is to log all accesses to these confidential files (most mainframe systems will allow the creation of file-opening logs). A few simple checks may be run on these logs in order to detect any unusual patterns of activity (such as use of the system outside normal hours or access from an unusual terminal); but very often, as with log-on logs, a quick scan by a responsible system supervisor once a day will identify many irregular accesses. Once a potential breach has been noticed, the main value of the log is to provide evidence for any subsequent investigation and to reinforce the confidentiality clauses in employees' contracts.

In most security situations, it is better to prevent a breach than to repair it, but when intellectual property has been stolen, it may be possible to prevent the thief from using or selling the information. Those responsible for computer security must also be aware of the law, which varies greatly from country to country in this respect. In some countries, theft of computer data is not even treated as theft, and so intellectual property held on a computer has very limited protection from the law.

Reference

[1]   Clark D.D., and D.L. Wilson, "A Comparison of Commercial and Military Security Policies," *Proc. IEEE Symposium on Security and Privacy*, April 1987.

Banking and Financial   14

14.1 SCOPE OF BANKING SERVICES

Banking and financial services cover a very large part of the whole economy. It can be argued that every other part of a market economy depends on them. Although the core business of banking has traditionally been defined as deposit-taking and lending, banks have for centuries offered other services, particularly in the area of payment systems and money transmission. Other companies offer more specialized services in closely related areas: insurance, pensions, share-dealing, foreign exchange.

Banking is now very dependent on computer technology, and all dealing and money transmission activities in particular rely heavily on computer networks. These networks have now become so central a part of the banking service that the companies who provide them are a fully integrated and critical part of the whole banking service.

This focuses attention on the security of these networks, both within banks and in companies using banking services. It is in financial networks that people contemplating fraud see the greatest opportunity for personal gain.

14.2 BANK ACCOUNTS

In the early days of computerized banking, there were many instances of bank staff creating new accounts for themselves (usually in a false name) and granting themselves credit. These would now be regarded as rather elementary breaches of procedure, and would be picked up immediately by internal audit and control systems. As with the commercial systems described in the previous chapter, however, it is almost impossible to prevent them occurring altogether, simply because staff must be able to perform these activities legitimately.

Account-opening checks should always include a unique identifier that can be *automatically and independently checked*: an identity card or driver's license number, company number, or tax or business registration number. Checks should also be made on duplicate accounts in the same name or against the same identifier, and in many countries there are now registers of false account applications that can be used in conjunction with credit references.

Once an account is opened, double-entry checks should provide the main line of defense against false transactions. Most transactions are automated enough that it is difficult to attack both halves of a transaction. A bigger threat comes from simple errors. This is particularly true for checks and other hand-written instruments, and is a strong motive for banks to move toward automated transactions in all spheres. The normal method of avoiding errors in capturing check data is double-keying: the data are entered by two separate operators and confirmed before being acted upon.

Procedural errors are also a problem. One area in which data communications can assist greatly is by making procedural help available on-line at the time that an unusual transaction is being carried out. Bank procedures manuals are very thick and are constantly changing; on-line manuals offer an ideal solution to this problem. CD-ROMs and data broadcast are both valuable tools in this context.

Bank internal networks are normally regarded as reasonably secure from outside intervention. They are generally private networks, using X.25 packet switching today, but moving in some cases to frame relay and asynchronous transfer mode.

14.3   ISSUES

14.3.1   Confidentiality

Confidentiality is often seen as a key issue in banking. It is a cornerstone of client relationships, although it is sometimes more difficult to define what people mean when they talk about confidentiality. Certainly, an outsider or bank customer should not be able to gain details of another's account, nor of account transactions that do not concern them.

Short of modeling client confidentiality formally [1], banks and others who provide financial services must go through the normal process of determining a need-to-know matrix for all data sets. One problem that may arise is that the need to know varies; a person may have a genuine need to know when carrying out a particular task, but should not have general access to the data at all times. In this case, the "agent" concept mentioned in Chapter 11 should be applied, so that the access rights depend on the person *and* the task.

Distribution of keys in a banking environment is also sensitive, since it implies a certain level of trust between banks. The ISO standard for key management [2] was designed with interbank applications in mind, but not all applications make use of it. Within a bank, extensive use is made of two-part keys, meaning that two employees must be present to effect a transaction.

14.3.2 Anonymity

Some customers want to be able to carry out some transactions anonymously and untraceably. While this requirement always raises doubts of legality, it arises remarkably often and is almost certainly not only for illegal transactions. Often the requirement is not for full anonymity, but only to ensure that the recipient of the funds does not know from whom they came. In other cases, anonymity is preserved to the point of using numbered accounts (although even the Austrian and Swiss banks who used to specialize in numbered accounts now require a demonstration that they are a legitimate requirement).

Complete anonymity, even at the transaction level, conflicts with the requirement for audit and control in a banking system and should be avoided whenever possible. What can be done, however, is to encode the data at an early stage, using a key that can be kept securely within the banks' systems and used only in audit checks.

14.3.3 Integrity

A stable modern society requires a dependable currency, banking, and money transmission system. When there is any suggestion that a part of the system is defective, it affects customer confidence and drives people away from the banking system. Small errors have a disproportionately large effect. This has arisen over "phantom withdrawals" from ATMs, counterfeit credit cards, and faulty retailer electronic terminals. While the data communications or computer procedures involved did not cause any of these problems, they could in all cases have been used to help avoid or alleviate them: a log of ATM withdrawals on a card could be made available to the customer, cards could incorporate higher levels of security check, and faulty terminals could be traced by analysis of transaction data across acquirers (the banks that handle transactions from retailers). As we will see in this chapter, progress is being made toward some of these goals.

Bland statements of confidence by banking representatives have little effect and may in fact be counterproductive. What is needed is proof of the integrity of the system, which can be provided by modern encryption and message authentication techniques.

14.3.4 Audit and Accountability

Closely linked with proof of integrity is the need to audit individual transactions. It is a fundamental requirement of most payment systems that every step on every transaction can be traced. Bearing in mind the requirement for confidentiality mentioned above, this may sometimes require encryption or encoding. It is most often met by providing a chain in which there is one common reference or a reference back to the previous stage.

There is a small but increasing demand from consumer groups for audit information to be available to customer representatives in cases of dispute. This raises a trickier set of conditions, and may be difficult to meet with current systems.

14.3.5   Disintermediation

Banking and financial services are increasingly electronic and anonymous, markets are more open, and information is ever more widespread. This raises the possibility of customers meeting their financial services needs without passing through the banks: companies raise money by the use of commercial paper rather than bank loans, international companies meet their foreign exchange and money market demands internally, without recourse to banks, and so on.

This process, which is called *disintermediation*, is a threat to banks. A further threat comes from the importance of data communications and networks in the banks' operations. Many of these networks are operated by outside companies. While many of them are in fact owned, or majority-owned, by banks, those that are not are in many cases seen as a threat to the banks. These threats are one reason why banks prefer to use data communications and security standards developed specially for use in and by the banking industry and not shared by outsiders.

14.3.6   Internationalization

Another major factor is the internationalization of banking and financial services. Demands that in years gone by would normally have been met by domestic markets are now often passed to the international markets. Companies may raise capital in many markets, buy and sell more abroad, and conduct operations on a European or global scale. Consumers travel more and have the same requirements for cash and spending in other countries as they do at home.

This internationalization not only leads to a need for international rates to be closely harmonized to avoid arbitrageurs making excessive profits at the expense of "normal" customers, but also requires a high level of international standardization and interworking of systems, so that money transfers and other operations can be conducted smoothly across borders. Given the highly controlled and rigid nature of many banking institutions, domestic standardization is difficult enough to achieve; international standardization threatens to take centuries. Some of the organizations that have sprung up to meet this international demand therefore tend to make their own rules and are seen as fickle by their own customers.

14.4 FINANCIAL MARKETS

14.4.1 Shares and Other Securities

Perhaps the most prominent of the financial markets are the stock markets. They are quite concentrated, and used to depend on *open outcry*—literally shouting matches on a single trading floor. Most stock markets are now turning to electronic dissemination of information and computerized trading.

There are two methods of operating a stock market: by *matching* buyers with sellers at a given price, or by having *market makers* who intervene between buyers and sellers and improve the liquidity of the market. Both depend for their fairness on everyone receiving the same information at the same time. Delays or loss of service affecting even one customer are therefore critical failures for the network. Most dealing members of a stock exchange will therefore have more than one route to access the information: data broadcast methods (by radio, television, or satellite) are increasingly popular as backups to a wired network. Large dealers will have four or five different information sources.

Reliability is the key criterion for the information dissemination; confidentiality is secondary, and relates mostly to subscription control. For the return messages, however, in which customers are dealing (buying and selling shares or options), both confidentiality and message authentication are critically important. Several exchanges currently make use of a key management system broadly in line with ISO 8732. Exchanges generally restrict dealing to registered members, however, and so some are considering new schemes whereby the exchange's central system can act as a trusted certification authority, issuing authentication certificates using the techniques described in Section 10.6.

14.4.2 Foreign Exchange and Money Markets

The foreign exchange and money (bonds and loans) markets also depend on rapid transmission of data to a large number of parties. They both trade enormous volumes, far larger than the underlying demand for foreign trade finance, and fortunes can be made by exploiting small differences in rates. Unlike a stock exchange, however, there is no organization to take control, set standards, or act as the trusted system.

Many banks and groups of banks have set up their own foreign exchange networks to overcome this problem. The information is again disseminated quite widely, and dealers who are not registered on a bank's network trade with it on the telephone. There are opportunities for fraud and errors here, but most banks that are active in foreign exchange have tight controls by means of individual dealing and exposure ceilings, which limit rather than eliminate the risk. There is scope for using EDI techniques in areas such as this, but it would be difficult for a third party to set up such an interbank system without causing an additional exposure risk.

14.4.3 Derivatives

The same principles are true to an even greater degree for derivatives trading. Derivatives are instruments whose value is linked to some other asset or instrument assumed to have a "real" value. Valuing them is complex and often requires elaborate mathematics; their details are frequently understood by quite a small number of people. Trading in derivatives is usually not automated, but the trades must be registered and controlled through the company's own network, which assesses the exposures and values the positions. There have been several instances where these control systems have been quite inadequate to the complexity of the task, resulting in major disputes over the value of positions and charges of fraud. The Kidder Peabody case mentioned in Chapter 3 and other recent cases involving Metallgesellschaft in New York and Barings in Singapore, are a few high-profile examples. In none of these cases was the data communication or network security directly to blame; the fault lay in the software that assessed the value of the position and allowed traders to build up unbalanced positions.

14.4.4 Commercial Paper

As we mentioned earlier, many companies now raise cash by issuing their own paper rather than passing through banks. Many of these issues are simply never traded, but an issue is most likely to be successful if a secondary market develops. As with the foreign exchange and money markets, those who trade in these areas make their own rules, and are most likely to be controlled only by their internal systems.

14.4.5 Financial Information Providers

Arising from all these activities is an enormous market for financial data, ranging from real-time stock market, commodities, and foreign exchange data to company news and broker comments. Giants of this market include Reuters and Dow Jones.

These services are often provided at relatively high prices to subscribers only. Security was traditionally maintained by providing the service on leased lines only, which ensured confidentiality (for closed user group services) and subscription control.

Analog leased lines are now generally considered too slow for top-end customers (although they are still perfectly adequate for many users). The alternatives include privately constructed fiber-optic links, multiplexed fiber-optic links through a private network, ISDN, and satellite networks. With the exception of the direct link, which is better than a leased line from every point of view, all of these have some security implications. Private fiber networks are often accepted as having adequate security for information provision; in the other cases, the data feed is likely to be encrypted for decryption in hardware or firmware at the receiving end.

14.5   INSURANCE

Insurance companies often run quite small branch networks, but with large numbers of salespeople. The branches are likely to be linked to a head office by packet-switched networks through a VAN operator. Much of the traffic is relatively undemanding, but on those occasions when rates change, quite large volumes of data may be transmitted, and these are regarded as sensitive by the insurance companies. This is a case in which the mere existence of a large traffic volume would be adequate to give useful information to a competitor; completely private networks are therefore a greater risk in this respect than a VAN, where the rate change may pass unnoticed in the overall traffic volume. The actual data transfer will be subject to normal controls, including MACs and possibly encryption.

The exchange of data with mobile salespeople is a different matter. Again, rate changes are sensitive, but even details of individual proposals and quotations are subject to strict data protection and consumer protection regulations. It is not uncommon for salespeople to print quotations using a briefcase computer and printer on a customer's premises. In some cases, quotations are even prepared on the central computer and transferred by modem to the salesperson's terminal. The public switched network must be used for this purpose; even videotex networks have been used. This requires a very good access control mechanism at the gateway to the insurance company system, and probably a "fire wall" between the part of the system used for producing these on-line quotations and the rest of the network.

14.6   MONEY TRANSFERS

14.6.1   Traditional Clearing

Figure 14.1 shows the traditional clearing mechanism used for payments between two trading partners. Able Baker Limited (AB) issues an instruction to its bank to transfer money to Charlie Dough and Company's (CD) account, which

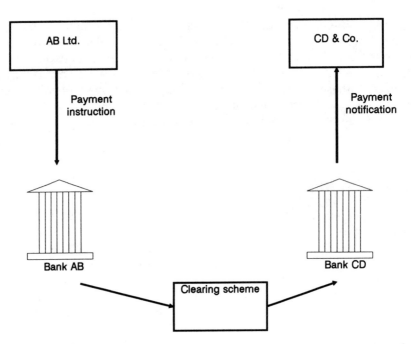

Figure 14.1 Traditional clearing.

is with a different bank. Or they may issue a check to Charlie Dough, who presents it to his bank.

In either case, the banks exchange this information through a clearing system: CD's bank deposits the money in a settlement account with the central bank, and AB's bank takes the credit and passes it to AB's account. The whole operation can take several days, since it still takes the time it used to take to transfer paper checks and other instruments. During at least part of this time, the money is in the banks' hands, earning interest, rather than in its customers' accounts.

Nowadays all clearings in the industrialized countries are electronic; the banks, or a central bank, run a dedicated private network for this purpose. These networks are limited to a very small number of banks—the clearing banks—in each country, and incorporate a very large number of security measures, including usually a requirement for two people, both with their own smart cards or other security devices, to be present to initiate an exchange.

14.6.2 Automated Clearing

In most industrialized countries, there is now a parallel clearing system (Figure 14.2) to which companies have direct access. It allows them to post transactions

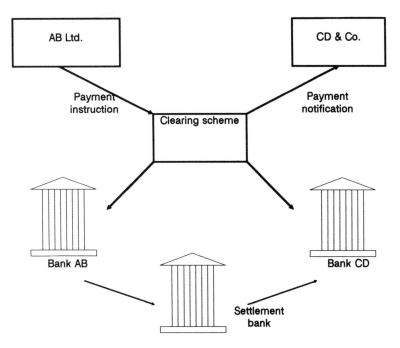

Figure 14.2   Automated clearing.

(whether debit or credit), which are then executed by the clearing system. The rules of the clearing house will determine what transactions can be set up by whom, how customers are sponsored into the system, and so on.

Security on these systems is at a lower level than for the central clearing, but there is nonetheless an important requirement for message authentication, which may be carried out entirely within the software provided by or for the clearing system or, as in the case of the U.K. BACS (Bankers' Automated Clearing Service), via a key-generating token issued to each customer. Customers are responsible for ensuring their own security below this level; the security provided by the system is equivalent to that on the company checkbook, with a limited number of authorized signatories.

14.6.3   International Clearing

The main international clearing house is the Society for Worldwide International Funds Transfers (SWIFT). SWIFT is recognizably an EDI service, with its own message formats and field types (special SWIFT-EDIFACT messages have also been defined). Only banks and securities houses have access to SWIFT, which has security equivalent to an interbank clearing system, though with a rather large number of participants.

Most trade payments are made through a system of *correspondent banks*—banks with whom the paying bank has a contractual arrangement. The correspondent banking system often involves long delays and high charges, and has come under strong attack in recent years, particularly from the European Union. There is an undoubted need for an international clearing system adapted to small- and medium-value trade payments. Some small networks exist by mutual arrangement, but they have failed to expand as fast as many had hoped.

14.7   FINANCIAL EDI AND ELECTRONIC BANKING

Companies who use EDI for sending orders, delivery instructions, and other commercial messages to their trading partners would often benefit from using EDI to communicate with their banks as well. Several banks have set up EDI systems to make this possible, but in practice usage remains low compared with trade EDI. One of the main reasons for this is that company treasurers feel that using EDI takes some of the control for making payments out of their hands, which should never be the case. Arrangements for authorizing and making payments are every bit as flexible with EDI as with checks or bank giro. The arguments in favor of EDI for invoicing and order transmission also apply to financial EDI: fewer errors and easier checking.

Financial EDI systems use the same standards as trade EDI: primarily EDIFACT in Europe or ANSI X.12 in the United States and Canada. The Royal Bank of Canada runs an EDI service that offers both trade and financial messages, thereby increasing usage. In Japan, where many EDI systems are closed user groups, financial EDI within the group is much more common. The systems used for financial EDI in Europe and the United States are generally separate from those used for trade EDI. They have a much more secure user interface, usually with a smart card or token as a part of the log-on sequence and a message authentication scheme operating during the transfer. Automated clearing, which has received much wider acceptance, is another form of financial EDI, but with proprietary protocols and message formats.

In some respects, consumers have been quicker to embrace electronic banking than companies. Security on consumer electronic banking is quite different. No hardware is normally involved, but special software, possibly including a proprietary cryptographic package and a multilevel password system, is typically used. Responsibility for password control remains with the user.

14.8   ELECTRONIC FUNDS TRANSFER AT THE POINT OF SALE

14.8.1   Retail Card Payments

The most widespread use of computer networks for transmission of payment information is electronic funds transfer at the point of sale (EFT-POS). EFT-POS is now almost synonymous with card payment in retail outlets, although there are a few other situations, such as paper vouchers, vending, and mail order, to which special conditions apply.

Card payments are growing rapidly in relation to other forms of payment such as cash and checks in every major country. Older forms of card acceptance, using paper vouchers, are dying out rapidly as they become fraud-prone and require too expensive an infrastructure. Card acceptance in most countries now requires an electronic terminal. This terminal may be specially designed for the EFT-POS application, with its own keypad, display, printer, card reader, and communications ports. The terminals are often owned by a bank or rented to the retailer by a bank. Or EFT-POS software may be included in the electronic till or electronic point of sale (EPOS) terminal belonging to the retailer.

In some countries, security was considered paramount in the design of the EFT-POS networks. In Denmark, for example, the banks commissioned a special encryption module design, which must be included in every terminal. Customer identification in these cases is usually through a PIN—a four-digit number which the customer must remember and which unlocks the card's payment functions. PIN verification is carried out by the bank's computer, and so every transaction must take place on-line. PINs should never be transmitted in clear-text (unencrypted) form; there is an ISO standard [3] for PIN encryption and PIN management which is usually followed.

In other countries, notably France and the United Kingdom, the aim was to replace the paper vouchers with electronic data capture. To encourage usage, costs had to be low, which meant allowing a large number of transactions to proceed off-line, with data collection overnight. High-value or suspicious transactions, and some others selected at random, would be sent for on-line authorization.

Some retailers now also hold a large "hot card" file, which can act as an additional reason to decline or authorize a transaction on-line. These lists may be downloaded overnight, or continuously by data broadcast. The risk management techniques used with the terminal can be controlled by the acquirer in the same way.

Payment cards first made their appearance in Europe as credit cards. Purchases were charged to a central credit account, which would then be cleared by the customer at a later date. With a debit card, which appeared later, the purchase is charged directly to the customer's current account (which may also have overdraft or credit facilities). It is sometimes difficult for customers to distinguish between the two, although in practice the majority of cards that can be used internationally are credit cards, and those that can only be used domestically are debit cards. Travel and entertainment cards (principally American Express and Diners Club) are a separate category, generally similar to a credit card but with different terms.

Most card payments, including virtually all credit card payments, take place under the rules of one of the three main card schemes: Visa, MasterCard, or Europay, using the pattern shown in Figure 14.3. Each of these has a network that is used for clearing transactions between members. Members are divided into two groups: those who issue cards to consumers and those who handle a merchant's transactions (acquirers). An acquirer must also be an issuer, but only a small proportion of issuers are acquirers.

Each country has its own standards for EFT-POS. The most common are the Visa and ISO 8583 standards, although they are frequently modified for local use. The U.K. Association for Payment Clearing Services (APACS) standards are also used in several other countries, while France and Spain have their own domestic standards. To the author's knowledge, only Denmark has a dedicated network for EFT-POS. In other countries, the public switched networks, both analog and digital, are used between retailers and acquirers.

Authorization is generally regarded as a low-security operation: under the APACS standard, all data are sent in clear. The ISO standard does not stipulate what encryption should be used, but does provide for a MAC for clearing messages. For on-line transactions and overnight collection of stored transactions, a MAC is normally applied. The MAC will include the transaction sequence number and, in most cases, a transaction key, which is given to the terminal by the acquirer host on completion of the previous transaction.

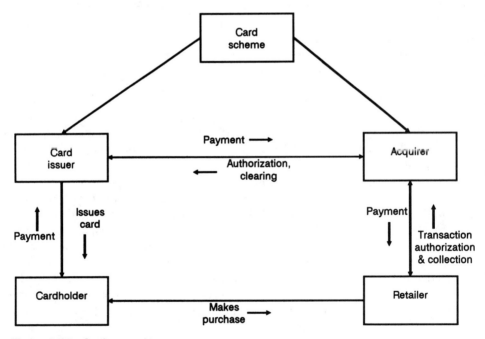

Figure 14.3 Card payments.

14.8.2 Personal Identification

Since the card is used as a proxy for the person in an EFT-POS transaction, the system is faced with two problems: how to prove that the card is genuine (a card authentication method (CAM)) and that the cardholder is the person to whom it was issued (a cardholder verification method (CVM)).

The traditional CVM is the signature, which is widely ignored by retail staff, because it is felt that handwriting is easily copied within the range of reasonable accuracy of a visual check. The most commonly used alternative (the only one practical in unattended situations) is a PIN.

PIN security has long been an issue: many customers cannot remember a four-digit number, particularly if they have two or three cards with different PINs, and so write the number on the card or keep the two together. It has also been feared that in some situations an unauthorized person could capture PINs electronically or by seeing them used. To date, however, no practical alternative has been found.

The alternatives proposed are mostly biometric: measuring a physical characteristic or behavior. The *physical* characteristics include fingerprints and thumbprints, finger length and hand geometry, and retina and iris scans. Most of these are quite good at avoiding false rejection, but the amount of data that must be held to ensure no false acceptance is rather large, and many of these biometrics are regarded as unacceptably invasive.

Behavioral characteristics include signature patterns (including pressure and speed as well as the resulting shape) and voice recognition. Since people's behavior changes from day to day, these techniques tend to require large amounts of data in order to avoid false rejections.

There has to be very widespread agreement on a biometric technique before anyone would invest in the technology on the scale required. This is likely to take a very long time. The banks have demanded a very high threshold for both false rejection and false acceptance. A combination of two techniques would make this much easier to achieve, but would double the investment.

Encryption could offer an alternative in the form of zero knowledge testing, which would be attractive in some situations. An example of this has been proposed by Feige, Fiat, and Shamir [4], although this method could only be used with smart cards and in attended environments.

14.8.3 Security and Audit

Other security issues surrounding payment cards include:

- What happens when a card is *reported lost or stolen*? It can be blacklisted ("hotlisted"), but this information needs to be available when the transaction is authorized. If the transaction takes place off-line, this means that it must be available at the terminal (implying large local storage and a broadcast technique). Even for on-line transactions, many transactions are authorized by the acquirer under delegated authority; acquirers must also have access to a comprehensive hotlist. In an international system, the management of hotlists is a major issue.

- *Fraudulent use* of the card, if it has not been reported lost or stolen, can often be detected from the pattern of use. In other cases, one of the CVMs mentioned above must be used. Both of these techniques are easier with the additional data storage afforded by a smart card.

- An increasing problem with magnetic stripe cards is *counterfeiting*. The equipment for printing, embossing, and encoding bank cards can now be purchased for under $5,000. The holograms are more difficult to duplicate, and most cards have some additional security mechanism built into the stripe which can at least be checked in an ATM. Nevertheless, a good counterfeit card will always pass muster at a retail checkout. Machine-readable watermarks and holomagnetics have been tried, but the most viable anticounterfeit measure is the smart card. The investment required to produce counterfeit smart cards, although not theoretically impossible, would be completely unrealistic when compared with the gains.

- The *distribution of PINs* is also a difficult issue. PINs are normally selected by the card issuer computer at random, and then sent to the customer by post (separately from the card). Some card schemes allow customers to change their PIN on-line (through an ATM)—customers can then choose a number that has some significance for them or that reduces the number of PINs they have to remember, and there is evidence that this reduces the likelihood of their writing the number down. Some cards have an algorithmically determined "offset" held on the card for PIN-checking purposes.

14.8.4   Smart Bank Cards

The first smart bank cards proposed were the Supersmart cards piloted by Visa. These cards had a miniature liquid crystal display (LCD) and keypad as well as the contacts, and local operations such as PIN input could be carried out directly on the card.

The French bank association Groupement Cartes Bancaires (CB) launched its first smart card pilot in Brittany in 1986, and by the end of 1993 all French bank cards were equipped with an integrated circuit. The standards for these cards are set by CB, and the cards include a pattern-of-use check as well as an internal PIN check. The results from this trial have been very good, and now almost 40% of fraud in France is on non-French cards.

The smart card will now be extended to other countries. Visa, Europay, and MasterCard have agreed on a common set of standards for a smart bank card [5], but are leaving the application details to the scheme and issuer. The card incorporates MAC generation and verification functions, as well as protected storage areas accessible only through the application program on the card.

14.9 Automated Teller Machines

All ATMs are operated on-line, and each individual network incorporates good security, with comprehensive network monitoring, MAC, and at least partial encryption (PINs, for example, are never transmitted in clear). There are potential weak points at the growing number of network interconnections, but security at these gateways is closely monitored using state-of-the-art host-to-host checks.

Mechanical failures of the machines are more common than electronic failures. From a network point of view, however, the shortage of comprehensive *audit* information is regarded as a weak point. Customers reasonably demand a higher level of proof for suspect transactions than is readily available today.

14.10 ELECTRONIC PURSES

A new area of smart card application is electronic purses. Smart cards have long been used as prepayment cards in single applications: for telephones or travel, for example. When the card can be used across a wider range of applications, however, it is treated as a purse containing electronic money.

Electronic purses can be either open (they can be used by anyone) or locked (for use only by its owner following input of a PIN). They are more attractive than credit or debit cards for small-value transactions, although this advantage can be canceled out if a comprehensive audit system is put in place.

A recent scheme called *Mondex*, developed by National Westminster Bank in the United Kingdom, gives holders of an electronic purse the opportunity to make person-to-person payments through a *wallet*, which resembles an electronic calculator, containing an equivalent smart card. This system is completely anonymous, although each transfer is secured by the smart card chips using a MAC.

The store of value in the Mondex system is held on an integrated circuit, which may be on a smart card or in the wallet. Value may be transferred between the two using a Mondex-designed MAC security handshake. Other features of Mondex include the ability to hold up to five currencies, loading through an ATM or special telephone, and optional locking by the customer.

References

[1] Brewer, D.F.C., and M.J. Nash, "The Chinese Wall Security Policy," *Proc. IEEE Symposium on Security and Privacy*, May 1989.

[2] ISO 8732, "Key Management to Achieve Security for Financial Institutions Engaged in Financial Transactions (Wholesale)," International Standards Organization, 1988.

[3] ISO 9654, "Banking—Personal Identification Number Management and Security," International Standards Organization, 1991.

[4] Feige, U., M. Fiat, and A. Shamir, "Zero Knowledge Proofs of Identity," *J. Cryptology*, 1988, 1(2).

[5] "Integrated Circuit Card Specifications for Payment Systems," Europay International SA, MasterCard International, Inc., Visa International Service Association, October 1994.

Subscription Services

15

15.1 ISSUES

Organizations that offer services to other organizations on a contractual basis, usually in return for a subscription or usage charge, face a number of additional issues in relation to their data communications security.

15.1.1 Subscription Control

How do they ensure that only registered users, or those who have paid their subscriptions, can gain access to the system? The basis of this control is normally a user ID and password combination: when the ID is offered, it is checked against a list and rejected if no longer valid. Or the password may be made invalid, so that the user account remains intact for billing and other purposes, but access requires a call to the system administrator or accounts department.

Very often, it is preferable to grant access to lapsed subscribers, perhaps to a limited set of material or for a limited time; they are then more likely to pay to continue their subscription than if summarily refused access. The "old" ID/password combination is simply given reduced access rights, or directed to a different start program that communicates directly with the subscription control program. All attempts by lapsed subscribers to log on, whether allowed or not, should be recorded.

Tokens or smart cards used for access can be given an expiration date, so that the token will not work after that point. Some encrypted services carry out subscription control by scrambling the user's key, so that he or she cannot decrypt the service.

15.1.2 Data Integrity

All data communications services are concerned with data integrity concerning both accidental and deliberate corruption of the data. For subscription services, however, there are additional commercial implications:

- Who is liable for any incorrect data? It is not always easy to trace the source of any corruption.
- Is it possible to flag any incorrect or suspect data (e.g., data that have failed a CRC check or were not updated at an appointed time)? Single-byte errors in a text string can usually be recovered by eye, but a single erroneous or missed digit can be catastrophic.
- What if the recipient takes action on the incorrect data and suffers a loss as a result? Or fails to take an action and loses a profitable trade? Can the service provider be liable for consequential damages?
- What, in fact, is the value of the data?

All of these points may be dealt with by contract, and it is generally important that they are. However, customer confidence in the service may be disproportionately affected if errors are detected.

15.1.3 Continuity of Service

For many subscribers, continuity of service is as important as its accuracy. A travel agent unable to access a booking system may lose a sale, a currency dealer who cannot obtain a dollar/deutsche Mark rate from one service will turn to another—probably for a very long time. People who use on-line services use them because they are there all the time when needed.

Many on-line services have a tendency to collapse just when they are most needed—under heavy load. It is critically important for these services to find a form of "graceful degradation": a slow and steady deterioration or reduced service is preferable to total collapse.

Encrypted services are particularly prone to catastrophic collapse under heavy loads, since there are more exchanges and acknowledgments per message, and time-outs often occur. It is possible to react to this by increasing time-out periods (some mainframe systems do this, but PCs and terminals attached to them rarely if ever do).

Fall-back systems often have less security than the main system, and the decision to invoke them should not be taken lightly. In some cases, it may be possible to bring the fall-back system into play to reduce the load before the main system collapses; users may then be unaware which of the two they are using, so that they cannot exploit the weakness. At times of heavy loading, for example, development tasks may be given a very low priority or are halted altogether.

15.1.4 Privacy and Legitimacy

Services may be provided on a subscription or *closed user group* basis for two main reasons: to derive income for the service provider from the subscriptions and to control access to the data or service. In the first case, the service provider is primarily concerned with subscription control: a legitimate user is one who has paid a current subscription. In the second case, other users, or the original data provider, may be much more concerned with checking that the data are only being supplied to someone entitled to receive them. As this implies, the service provider does not always own the original data.

In both cases, it is more difficult to prove that the person or entity originating an exchange is legitimate than to prove the legitimacy of the registered subscriber. IDs and passwords, as we have already discussed, may be copied or stored in communications programs.

Systems concerned with this aspect will not only insist on frequent changes in passwords, but may also require human rather than automatic input of the password (this is easily checked by the speed and irregularity of the human operator compared with a program). In extreme cases, further precautions may be taken, such as issuing smart cards or tokens.

A further opportunity is afforded in those countries and systems that can support calling line identification (CLI); that is, the telephone network can show (with some limitations) the identity of the number from which the call originated. Whereas a subscriber may have legitimate reasons for using a different number (e.g., when working at home in the evening), security-conscious systems could require any alternate numbers to be registered.

15.1.5 Logging

While other data communications systems have a requirement to prove the nonrepudiability of a transaction, subscription services will normally log all activities. As a minimum, every log-on attempt must be logged, as well as every log-off and any access or attempted access to a gateway. Some systems will also record precisely what data were accessed, or at least any access to sensitive or unusual data.

It is important that these logs are reviewed manually *and* automatically analyzed. A failure to analyze logs can allow small problems to become serious, and there have been many instances when a sharp-eyed operator has detected attempted security breaches and been able to prevent an actual breach. Manual review will usually show up any obvious problems, but automatic analysis has the advantage that it can often spot exceptions more quickly, as well as pick up more obscure patterns of unusual behavior.

The accounting suites of many subscription services provide a useful additional logging service. Because these reports are passed to the customer, any security breaches or unusual patterns on that side can also be detected—if they are analyzed rather than passed directly to the accounts department.

15.1.6 Control Points

With a subscription service, there is usually one extra control point in the system: the initial registration. This provides an important facility, and the marketing department's keenness to secure a new subscriber should not override the need for security, payment, and, if necessary, credit checks. Nor, however, should the checks be so onerous that they put the potential subscriber off completely!

It is often useful to record the information given on the registration form and to check that it is consistent with information gained on subsequent accesses. For example, a CLI may provide proof that a person is indeed calling from the number given. Or a sudden interest in nuclear physics from a taxidermist's shop might be regarded as unusual.

The next control point is a log-on attempt. A new subscriber's first log-on should always be treated as an exception, and may be subjected to additional controls or checks.

Accessing a gateway from within a service should form a control point. "Invisible" gateways may also be set up within a database to provide additional control points for sensitive data or simply to provide a more complete log.

Control points may also be time-based: subscription services are often charged on the basis of connect time. Very long sessions in particular should be subjected to additional scrutiny.

15.1.7   Payment

On-line networks provide many opportunities for companies to advertise goods and services. They also want to be able to capitalize on the advertisement by making it as easy as possible to make the purchase. The easiest way is clearly to fill in a form on the screen. Some services, such as an on-line credit reference agency or a financial market information provider (IP), will also require payment before making data available. They can require subscribers to register by post and set up an account, but this destroys one of the greatest benefits of being on-line: immediacy.

Making payments on-line is less easy. On a small commercial VAN with a limited number of subscribers, the VAN provider may be prepared to add some charges to the customer's invoice or to provide the customer's billing details to the supplier (this would normally require the customer's explicit permission). But on a large-scale system, neither party is likely to be prepared to accept the delay or payment risk. Some way must therefore be found to make payments through the network. Direct debiting using an automated clearing house, as described in the previous chapter, would be possible, but it is very unlikely that any bank would consider setting this up until security can be shown to be sufficiently high. Credit cards provide the obvious intermediate stage.

Several companies have set up schemes for making payments on the Internet:

- A scheme that uses preregistration of credit card details and open email to confirm transactions reported by vendors before charging items to a credit card account;

- An encryption system (based on the PGP algorithm discussed in Chapter 10) that allows credit card details to be sent by email (both parties must have copies of the package);
- An electronic money scheme, still at the pilot stage, that involves the exchange of secure messages between trusted systems in a manner similar to the Mondex system (also discussed in Chapter 14).

At this stage, however, no route has been widely enough accepted for any kind of standard to be formed, and it seems likely that trading of goods and services (as against sales of data) on subscription networks will remain restricted for at least another year or two.

15.1.8 Fire Walls

Customers of a subscription service may reasonably insist that the network protect or indemnify them against harmful effects such as other users hacking into their systems either by making use of channels intended for outward access or by gaining access to legitimate users' passwords and access rights. If the network operator's guarantee or indemnity terms are not adequate, a subscriber must erect his or her own "fire wall" to protect the local system. There are two main forms of fire wall: independent systems and pattern-of-use monitors.

Independent systems, most often PCs, may be used to gain access to the remote system. The PC should be logged off the local network while it is in use on the remote network.

Pattern-of-use monitors act as a filter, passing on transparently all allowed traffic and intercepting any suspect traffic for confirmation before it is allowed through. Such systems can only operate effectively on rather fast hardware that is capable of handling at least twice the "real" data rate. Currently available packages can be very effective for predictable traffic flows in large commercial organizations, but are less likely to work in a more fluid research or engineering environment.

15.1.9 Intellectual Property

Protection of intellectual property rights in a computer environment is always tricky. On a subscription network, the problem is compounded by the difficulty of establishing ownership of the data. Very often the network service provider is not the IP.

Contract terms usually give the user the right to use the data internally but not to sell them in any form. Most networks and access systems, however, allow the subscriber to download large quantities of data. If the subscriber manipulates the data, possibly so far that the originals cannot be deduced from the derived data, then the situation is less clear and it becomes difficult to police the copyright. Information providers for whom this is an important issue must either be very explicit in a contract about the copyright of downloaded data and how it may be policed, or they must provide a software package that alone can decode or decrypt the data for display. There is probably no mechanism for actually preventing the data from being copied, and so the next best thing is to prevent it from being used.

15.1.10   Cross-Border Issues

When networks cross borders, more problems can arise. Several countries explicitly forbid the exportation of personal data on-line or in machine-readable form. The United States attempts to control the exportation of encryption software. Value-added services performed over a network are likely to be taxable and may even be subject to duty and customs controls.

Since it is clearly impossible for customs staff to control these aspects physically, they are increasingly careful when scrutinizing accounts in relation to services provided over international networks. Although this is not yet a major problem, users and providers of network services must be aware that the burden of proof may ultimately fall on them.

15.2   COMMERCIAL VANs

The best known commercial VANs are the multivendor access services: CompuServe, America Online, AT&T Easylink, IBM Information Network, and Datastar are amongst the most widespread today. These networks may provide some services using their own data, but the majority of usage comes from subscribers accessing IPs on the network, most often through a gateway.

Subscribers have an account with the service provider, and in some cases must also have an account with the IP before accessing the IP's gateway. Security on these systems depends mostly on an initial exchange of ID and password, although some IPs have more stringent checks at the gateway.

In Europe, another important set of VANs are the national videotex systems (e.g., Minitel, Prestel, Btx, Datavisionen). These low-speed but highly economical services still provide access to many useful sources of data. Security is low, and demonstration account passwords are often given away freely. There are now few security-conscious systems attached to the videotex networks.

The most widespread access service of all is the Internet, which is the exception to almost every rule in the data communications book. It has its own protocol, but this is so widespread that it is a de facto standard. It has no central administration, and there is little or no control over the data that is transmitted over the network. But it connects over 2 million computers and 30 million users.

IPs are the owners of the data or services (such as a credit-checking service or insurance quotation system) that can be accessed through one of these access services. The responsibility for ensuring that data are not freely copied or made available to the wrong people lies with the IP, who may in turn be answerable to the true source of the data (e.g., a stock market or insurance company). The tools at the IP's disposal have already been discussed; they include extra password checks at the gateway and special software given to subscribers to allow them to decode or decrypt the data.

15.3 CLOSED USER GROUPS

In a category similar to subscription services are closed user groups. Here entry is controlled not to raise revenue but to restrict access. For example, a car parts or electronics distributor may use a closed user group service on a commercial network to allow dealers to place orders or check technical details. Or a trade association may offer specialized information services to its members. This type of arrangement is very common in Japan, where many commercial VANs are run by companies for their dealers and subsidiaries.

Some commercial services may be offered only to a closed user group, while others may be offered to the general public or to all network subscribers. The key difference is usually payment: how is payment to be accomplished?

Within many closed user groups, a high level of trust is necessary against a background of relatively insecure communications: if a travel agent books an expensive holiday without the necessary authority, it may be difficult for the travel operator to prove that the transaction took place. Such systems are slowly being upgraded to include MACs, digital signatures, and other nonrepudiation checks.

15.4 ELECTRONIC MAIL

One service frequently offered on commercial VANs is email, which allows users to send messages to each other without their having to be connected at the same time. Several standards have been developed to promote this type of service, which offers major efficiency gains to regular computer users. The most important are the ANSI X.400 Message Handling Services and X.500 Directory Services standards.

Key factors in the success of an email system are privacy and data integrity. Email usually allows any type of data to be attached to the message, including binary files. Everything must be checked using a suitably long CRC; most standard email systems do this adequately. Privacy requires some enhancements; the most common standard for this is PEM, although some network providers have their own proprietary enhancements.

Email is also subject to some abuse: in particular the problems of junk mail and overuse of group addressing. These can lead to mailboxes becoming full of mail that is not worth reading and lessen the impact of valid messages. Making directories secret to some extent defeats the purpose, and so the only satisfactory answer, as well as good discipline on the network, is a rapid review feature on the mail package.

15.5   ON-LINE SERVICES

Some services are large enough or security-conscious enough to operate without recourse to shared-access networks. They mostly offer higher value services, where the increased cost of the communications link is less of a problem. The most important of these services are the financial market networks: in addition to the closed user groups operated by the major markets themselves, services such as Reuters, Dow Jones, and Extel provide financial market news in real time on a subscription basis. The core service for all of these is only available using a leased line or dedicated fiber-optic link, although all have or are considering a lower level of service that might be available on a dial-up or broadcast basis. The news wire services (e.g., Associated Press and Agence France Presse) also operate mainly using leased lines and privately owned microwave circuits.

Many database services can be accessed either directly or through one of the common access networks; these include legal, scientific reference, and bibliographic databases. One of the most extensive types of data available on-line consists of bibliographies and abstracts relevant to particular topics. Some of these are subject to preregistration for interest group purposes only, but the information is generally available without charge or at very low cost and is not encrypted or protected in any way.

Other services available on-line by subscription include:

- General information services (offering, for example, news, weather, train times, local events) usually provide the data free but may charge for connection time. Under the condition that the call can be charged to a registered subscriber, there are few security implications. Some of these services may, however, act as access routes into more sensitive services (such as the royal family ex-directory numbers found by a journalist through the U.K. Prestel system). In these cases, there *must* be extra controls at the gateway. Because these services are extremely easy to access, there must also be checks against casual abuse (e.g., timed log-outs after a period of inactivity or possibly even regardless of activity).
- Some services essentially consist of advertising, paid for by the advertiser, the caller, or both. These include contract opportunities, goods for sale, and services on offer. Advertisers want to have their material widely read, but

may charge or want to know who has viewed their material; security as such is not an issue.

- There are several categories of bulletin boards. The most common are those provided by companies as a help desk or for comments, while others are run by enthusiastic amateurs who recover their costs by using premium-rate numbers and low-speed lines. Many bulletin boards exist largely for the exchange of software and data files, and so reliable file transfer software is very important. Most bulletin boards use the very common shareware protocols, such as YModem, rather than one of the more efficient but proprietary protocols using techniques such as forward error correction and "sliding windows" (which make better use of full duplex lines).

- Entertainment and games offered on-line can represent large commercial opportunities. Again, the choice of an efficient protocol is important. Interactive games can require rapid response to individual keystrokes, and so the main backbone networks (which are usually packet-based) are extremely inefficient. Direct access by a high-speed analog modem is preferred, preferably with any error correction disabled.

15.6 BROADCAST SERVICES

A special category of subscription service, which we also mentioned in an earlier chapter, is that offered by broadcast media, usually satellite or VBI data broadcast. These services include several financial data companies, such as Knight Ridder and Bloombergs in the United States, Topic and Market Eye in the United Kingdom, Aktievisionen in Sweden. All of these companies rely on providing their own software (and sometimes hardware) at the receiving end to decode the data sent. A relatively low level of encryption or encoding is used on the data, since the transmission also includes control codes that instruct the software what data to store and display. This facility allows a very subtle approach toward overdue or lapsed subscribers, and can avoid the need for extensive downloads when the overdue subscriber does pay.

Several broadcast services are related to security, transmitting, for example, details of lost and stolen goods or credit cards. A moderate level of encryption is necessary on these transmissions to keep thieves from gaining more information about the lost items or knowing just when a stolen credit card goes "hot." Again, this is handled by proprietary algorithms and dedicated software in the receiving terminal.

Services offering items for sale or wanted generally have low security. They can rely on the mechanisms inherent in the transmission protocols and in the need for subscribers to have suitable equipment.

The most common broadcast data service of all is teletext, which is almost universal on broadcast television services in Europe, whether direct or satellite. In other parts of the world, it is also available but less widely used.

Most teletext services are unencrypted, although, as we mentioned in Chapter 12, they should be subject to the same encryption mechanisms as the analog television signals. The Videocrypt system uses the VBI (the teletext lines) for its own control signals and is therefore unable to encrypt these lines.

Other Applications **16**

16.1 SOFTWARE DISTRIBUTION

Most packaged software is sold on CD-ROM, floppy disk, or, for larger systems, magnetic tape or cartridge disk. These items are produced by bulk copying, and their readability is not always guaranteed. It is also not uncommon, particularly for upgrades, for some files to be missing from the distribution pack. Purchasers should insist that vendors warrant that their packages are complete, readable, and free from viruses, and should know what liability the vendor is prepared to accept for breaches of this warranty. This is clearly likely to be less for a $50 package than for a $100,000 customized suite.

Software vendors are increasingly making use of data communications for transmitting upgrades through bulletin boards for popular packages, or through a dedicated diagnostic link when the vendor provides support. In these cases, readability is no longer an issue, but it is doubly important to ensure completeness and freedom from viruses. A further issue in these cases is to ensure that the system has adequate backups of both the new software and any original software or configuration files. A high-speed error correcting modem or digital (X.25 or ISDN) link should be used whenever possible, and all files should be associated with adequately long check sums to ensure that virtually any error can be detected.

Any system that downloads files, whether from a regular source such as a head office or from a bulletin board, should have a competent virus-checking suite. Any executable file is regarded as a potential source of viruses. The number of files downloaded should be checked and any virus signature detection program should be run before any new software is executed.

These constraints should be applied to internally written and distributed software as well as to brought-in software. Typically, different people will be responsible for each part of the installation, and so a useful check will be established.

Another aspect to check with software distribution is the control of copying and copyrights. When a piece of software is subject to copyright, it cannot be copied freely, even within an organization; separate or multiple licenses must be bought. Many companies now use networks where the workstations do not have the ability to load software. In other cases, the network is run in a way that positively encourages the existence of multiple copies.

When workstations do have the ability to load or copy software, the only real solution is to hold and maintain a software audit. This is not just a policing exercise; it will often show up issues such as conflicting or out-of-date versions of software, incompatible configuration settings, or redundant software. Once done, the software audit is relatively easy to maintain, since only changes need to be looked at.

The other side of the copyright argument is that it is necessary to have adequate numbers of software copies, not only for backup purposes but also for the level of use within the company. If people who need access to software cannot obtain it legitimately, then they will resort to the easy step of copying it.

16.2   TELEMETRY MONITORING AND CONTROL

One of the most important uses of data communications in terms of volume, and yet relatively invisible to most computer users, is in telemetry: the transmission of values from instruments (flow meters, thermometers, pressure gauges), switches, and other remote equipment (even television monitoring cameras) from remote sites to a central site.

Telemetry takes place within a site (e.g., a factory complex) or across many sites (e.g., a water or gas distribution network). In the first case, a LAN is likely to be used, probably with relatively high bandwidth (1 Mbps or higher). The second is more likely to rely on a WAN operating at 64 kbps or lower.

Key applications for telemetry are in public utilities for control of electricity, water, oil, and gas networks, and in railways and other transportation systems. Rail networks and electricity distribution systems nowadays lay their own communications networks (typically using fiber-optic cable) alongside their track or electrical cable. Utilities in many of the countries whose telecommunications are being deregulated have turned their need for a secure network for their own purposes into an opportunity to provide commercial services.

In water, oil, and gas utilities, much lower speed networks are often used, with a mixture of low-speed modems, radio communications, and other media, depending on the environment and distances involved.

While normal error checking (block CRC and parity checks) can take place on the transmission in these cases, the key tests are in fact logical and consistency checks. When modems are used, even the built-in error correction standards (V.42 or equivalent) are not always helpful; the software may prefer to handle its own error detection and correction. Values in many of these environments change only slowly and in fairly predictable ways. If a "rogue" reading is spotted, the first step is often to call for a fresh set of readings; only if this confirms the first result is some action taken.

When the equipment, such as a piece of electrical switch gear, a valve on a pipeline, or a railway signal, is directly controlled from the central site, the need for complete integrity on the data communications link is even more acute. Such transmissions are often duplicated, and the remote equipment checks that the two transmissions are identical and that they both fall within expected limits before it takes any action. The new control values are often given as *set points* or target values, with the actual control being effected by the remote equipment to reach that value.

A railway signal or traffic-light controller may be sent a schedule of changes some time in advance. These are conservatively set to ensure a large margin for error, but can be speeded up by a subsequent transmission when, for example, the previous train has cleared the next section. In this way, the safety of the system is not dependent on any given transmission succeeding within a fixed period of time.

16.3 GOVERNMENT INFORMATION DISTRIBUTION

Governments worldwide are usually slow to accept technological innovation, even as they promote it in speeches and legislation. Forms must still be completed in triplicate and files are moved round the country in fleets of vans, while similar operations in commercial enterprises are highly computerized.

There are several reasons for this: in addition to the innate conservatism of civil and public servants, there is the sheer scale of investment and the number of interconnected systems that must move together when any change is made. Governments are also concerned about the privacy of personal data, commercial confidentiality, and the mileage newspapers can extract from any small leak.

Government networks are therefore kept largely separate from private networks (which also partly explains their slow growth). Throughout Europe and North America, these networks insist on a very high level of standardization, including full compliance with OSI and ISO standards. Partly driven by a small number of applications where there is a genuine need for the highest levels of security, governments have been in the vanguard when establishing standards in this area, such as the European ITSEC criteria.

There is thus no single rule as to the type of security protection used to maintain the confidentiality of the information distributed, but proprietary or nonstandard implementations are likely to be frowned upon or disallowed. Companies communicating on a regular basis with government departments are likely to have to implement several security standards and to ensure that their equipment and software continues to meet international standards whenever possible.

16.4   MILITARY AND NATIONAL SECURITY

This book is aimed at the vast majority of commercial users of data communications whose requirements for security are related to their business needs. The need for security or the potential cost of a breach must be balanced against the cost of the countermeasures.

In the case of military, police, and national security, purely economic arguments can no longer be applied. While economic factors are still be taken into account (all government departments have to work within a budget and their own internal priorities), the techniques used for setting priorities and for resolving apparently incompatible criteria such as the value of human life or public confidence are completely different from the criteria discussed in this book.

National security is an umbrella term which is sometimes abused. Traditional security classifications are a blunt weapon and would be greatly refined in a data communications or data processing environment. The term covers discussions between a finance minister and central bank governor on interest rates, a civil servant's briefing to a cabinet secretary on farm pricing, or a proposed announcement on child welfare. The need for highly secure messaging, with several grades and categories of clearance, is nowhere clearer than in government, and such a system could undoubtedly be implemented using current technology.

Most other applications of data communications within government have already been described and discussed in earlier chapters. As with all security, people are likely to be the weak link—the more important it is to maintain security, the greater will be someone's motive to break it.

Military command and control is perhaps the pinnacle of the security technologist's art. Again, the primary requirement is for highly secure messaging, but radio is the main form of communication used, which brings the problems of fading, interception, and jamming to the fore. A wide variety of spread spectrum techniques are used in addition to other forms of encryption, message authentication, and specialized key management tools.

Communication with satellites and missiles can be even more critical. As with many military applications, it is desirable to avoid disclosing the existence or location of a unit or a person.

The essence of encryption lies in using the most advanced tools and the fastest hardware available: even with full knowledge of the algorithms used, decryption without knowledge of the keys is several orders of magnitude slower. The problem thus resolves to one of key management, which is developing into a discipline in its own right.

Summary and Conclusions 17

17.1 FURTHER SOURCES

In this book we have attempted to give an overview of computer network security as it affects the majority of businesses, and from a pragmatic rather than a theoretical viewpoint. We recognize that not everyone who has responsibility for the security of a data communications system is a computer specialist or mathematician, while the programmers and computer operations staff who have to put security measures in place on the system do not always have the necessary broad view of the factors that may be involved.

Some readers will need to refer to more specialized books on cryptology, software techniques, or risk management, and a selection of these is mentioned in the Bibliography, located at the back of this book. We have also given references throughout the book for the main standards used.

17.2 ASPECTS OF SECURITY

We have included in our definition of security a wide variety of causes of failure: human or electronic, accidental or deliberate, short-term or permanent. These blunders, crimes, glitches, and crashes can all be very harmful to a business, but they are not always given equal weight by those responsible for the security of the computer network.

Personal safety is not normally considered in this area, although companies have responsibility for their employees' health and safety as well as security. We are mostly concerned with the effects loss of data, erroneous data, or disclosure of data have on a business.

The main criteria for data communications security are:

- Nondelivery: This is the chance that the data are simply not received. If detected, this is much less of a problem than if it goes undetected.

- Accuracy: Random errors can nearly always be detected and, with most forms of network, can easily be corrected.
- Integrity: In many applications, it is important to prove that the data cannot have been altered since it left the originator. Cryptography offers a possible solution here in the form of MACs.
- Confidentiality: Although many types of data are not confidential, there are other cases in which disclosure may be harmful. In these cases, encryption or simpler encoding mechanisms should be used.
- Impersonation and repudiation: Digital signatures can be used to prevent a person from denying that he or she sent or received a message, and should also prevent impersonation or masquerading.

Table 17.1 shows the main protection methods used against each of these criteria.

With these criteria and the knowledge of the damage likely to be caused by a security breach, users can draw up a *requirements specification* in terms of the minimum MTBIs for each type of failure.

Table 17.1
Security Criteria and Protection Methods

Criteria	Protection
Nondelivery	Watchdogs
	Message numbering
	Inclusion of message number in MAC
Accuracy	Parity checks
	Check sums and CRCs
	MACs
Data integrity	MACs
	Message encryption
Confidentiality	Coded fields (requiring separate lookup table or special software)
	Message encryption
Repudiation	Digital signatures
	Digital signatures on acknowledgments

17.3 PRECONDITIONS

A data communications network cannot be made secure in an insecure environment. Among the aspects that need to be considered are:

- The system design must reflect the current business process and not include too many redundant or unnecessary steps. Where possible, links between systems should be direct and not involve rekeying. Good ergonomic design reduces errors and the opportunities for casual security breaches. Systems should not be overcomplex.

- Physical access controls must cover all relevant equipment, not just user terminals. It is best if there is nothing exposed to view that might provide temptation.
- Employees will only behave in a trustworthy way if they themselves are trusted and if senior management sets a good example. An important opportunity to control staff is at the time of recruitment.
- Contracts for computer and network services must reflect the organization's requirements for security.

17.4 RISK MANAGEMENT

In order to establish quantitative criteria to determine which measures should be put in place, we need to establish:

- The object threatened: This is usually a file or an item of data.
- The grade of risk: How much would it cost us if the object were in fact lost or disclosed?
- The level of threat: How likely is it that there will be a problem? To be really sure that the threat is negligible, we would have to use rigorous analysis and proofs.
- The cost, legal, and time constraints involved in any solution: It would be unusual for there to be no technically feasible solution, although expense is often a limiting factor.

There are techniques and models that allow you to use these parameters to find an optimum solution, which will often involve a layered approach to security, with increasing security applied to more privileged functions.

Standards for computer security include the European ITSEC and American TCSEC evaluation criteria, which allow hardware and software tools to be tested against specific security objectives, as well as a number of standards in specific areas published by bodies such as the ITU, ANSI, and CEN.

17.5 HARDWARE, SOFTWARE, AND NETWORKS

Hardware items can be stolen, powered off, disconnected, tampered with, or simply used to monitor signals or traffic. Small and common items are the most vulnerable: modems, PCs, and monitors in particular. The design and layout of buildings and telecommunications areas also have a major effect on security.

Software problems can stem from the specification, design, testing, or run-time environment. Suitable tools and languages must be chosen when the implementation is inhouse.

Operating systems are a particularly important differentiating factor: some are simply unsuitable for any secure application. But it is also important to ensure that any packaged software is suitable for the environment, that application software is adequately specified, and that no unauthorized software is allowed on the system.

The OSI reference model proposes certain security services at different levels, but is not specific when it comes to the choice of tools to provide those services.

Databases are particularly vulnerable. Techniques for protecting them include encryption, comprehensive logging, or simply removing disks. A database administrator should be appointed.

Redundancy, recovery, and backup procedures must be designed into each system so that it can recover from faults when they do occur.

Networks and external communications systems are much more reliable than they used to be. But there are still big differences between the best and the worst, and it is important to pick the best network for your purpose.

17.6   APPLICATION-SPECIFIC RISKS

Real-time control systems use special techniques such as limit checks, voting systems, and fail to safety, which could be used more often in commercial systems.

Banking and financial systems make use of private networks and special security devices, but they still depend to a large extent on trust to maintain the confidentiality and internal controls that are essential in this environment. There would be opportunities to close the loop if more use were made of EDI for financial transactions.

Internal accounting systems and databases need to be set up so that users can access the data they require in their day-to-day work, but they should not have unlimited access. Customer databases are particularly prone to theft by employees leaving their company.

Although computer data are still regarded with some suspicion by many lawyers, contract details are often highly confidential. Data communications could often be used to ensure contract compliance, but are rarely specified. Intellectual property held on computers can be at risk. It is useful to identify sensitive files and to protect them by incorporating a date and time in their MAC fields.

Users of subscription services need to protect themselves against theft of their data, hacking, and abuse by authorized users. Particular care is needed with anarchic networks such as the Internet. Personal data on any network are likely to be subject to data protection legislation.

Radio and satellite-based networks have special problems of fading, crosstalk, interference, and monitoring. They are, however, widely used in military and national security applications, where national governments are concerned about the erosion of their special position in cryptography.

17.7 ENCRYPTION AND KEY MANAGEMENT

Many forms of protection involve cryptography: transforming the data using an algorithm and a key. The algorithm is not usually considered secret, and indeed there are several published algorithms. But the keys must be kept secret, and key management is an important part of encryption technology.

Keys may be symmetrical, as in the widely used DES, or asymmetrical (public and secret keys), as in the RSA scheme. Symmetrical keys are most often used for encrypting the body of the data and for MACs which ensure the integrity of the data transmitted.

Asymmetrical keys, which require more processing power particularly for decryption, are most useful for digital signatures. They are increasingly used in data communications, since they can be used to demonstrate the identity of the sender (and in the case of an acknowledgment the receiver) as well as the integrity of the data transmitted.

Key management poses its own problems, and often several layers of key are required. Some systems use data-encrypting and key-encrypting keys, while others use multipart keys, primary and session keys, or transaction keys and seed keys. A stack of keys may be used together with a pointer. Key management is best handled by an established software package from a reputable vendor.

New encryption schemes are now needed, but no scheme has yet achieved wide enough acceptance to replace DES or RSA in their respective fields.

In many situations, there is a particular need for an automatic identification test to replace the password, signature, or PIN. None of the biometric schemes so far advanced can meet realistic levels of false acceptance and false rejection simultaneously.

17.8 HARDWARE TOOLS

Hardware security tools operate faster than software and it is easy to see when they are in place. They include:

- Encryption chips for standard algorithms;
- More complex ASICs for specific cryptographic tasks such as key management;

- Secure bootstrap PROMs for PCs and similar systems;
- Dongles, most often used for protection of software copyrights;
- Smart cards, which can perform a wide range of identification and cryptographic functions, such as PIN verification or storage of a fingerprint or signature pattern;
- Tokens and special calculators used for key generation;
- Inline encryptors and encrypting modems;
- PC security modules, which can perform key management and additional security functions not provided by the operating system;
- Security FEPs for use on networks with midrange and larger systems;
- Tempest hardware, which is protected against electromagnetic radiation.

The last two are likely to be used only on very security-conscious installations, but all of the other devices are also available at moderate cost even for quite small systems.

17.9   SOFTWARE TOOLS

Software tools are usually less expensive to implement than hardware, and they can operate across a network or other data communications link. They most commonly address the problems of access control, user authentication, and file protection. Within the first category fall systems for ID and password entry on multiuser systems, boot control on PCs and other highly distributed systems, and access rights setting. Stand-alone file protection systems usually involve encryption of selected files. In all these cases, there is a problem in that there has to be a way of bypassing the protection in the event of a fault (since otherwise the legitimate user can be permanently unable to retrieve files), but this bypass route then becomes a weakness.

Software is also available for detection of viruses. Such software can never be absolutely reliable, but it is a necessary precaution in many situations; it needs to be run when new software is loaded and regularly thereafter.

Operating systems on small systems are generally insecure, but on large systems they can be made very secure. It is important for the person responsible for security to know the weaknesses of the specific operating system in use. There are specific tools available to improve the security of networks when designing and programming systems, as well as software to help automate the difficult process of key management.

It is also very useful for administrators to understand the tools and techniques used by hackers.

17.10   ACCESS RIGHTS MANAGEMENT

The supervisor must establish who can do what to what data. The limits of the rights he or she can grant or deny are set by the operating system and may be too coarse for some purposes. In these cases, access rights must be managed directly by the application. Access rights can be managed by limiting physical access, by controlling entry to tasks on a session basis (theoretically through a separate application), or by the application itself. Although the session controls are usually rather coarse, many standard applications do not have any controls at all, and so they are the only practical route. When a network permits indirect access to data, through a gateway, for example, application-level controls are almost essential, unless a system of delegated rights can be implemented.

17.11 TYPES OF NETWORKS

Many PCs within an organization are connected using a LAN. LANs can be operated peer to peer or can be server-based. Peer-to-peer networks are difficult to manage from a security point of view, since they depend on each user granting access to facilities.

WANs, MANs, and other extended networks normally provide certain key security services, such as confirmation of delivery, provision of source address, and connection confidentiality. But most encryption, authentication, and access rights management functions must be performed by application-level functions.

A company's internal networks must be controlled by limiting access rights without limiting people's ability to do their job; this often means careful thought about the structure of data and directories. Field sales networks are particularly vulnerable, and token-based log-on sequences are recommended when the data is sensitive.

Communication between companies is still often by magnetic tape or floppy disks. In many sectors, however, the use of EDI is growing. The EDI standards include a number of security services, but many existing services, operating through a VAN and central system, rely on a high level of trust. The special characteristics of wireless and satellite-based networks are usually addressed within the technology, so that users do not have to implement their own additional security measures.

17.12 APPLICATIONS

Administrators setting up access control schemes for a particular application must take into account not only the company structure, job definitions, and management controls required, but also any characteristics of the application that make it particularly prone to data communication or data entry errors, fraud, or other deliberate abuse.

For example, customer databases are of considerable value to a competitor and are often copied by sales staff leaving the company. Payroll and personnel records are often the subject of casual prying (but rarely of material damage to the company). Salespeople's bonuses are particularly sensitive. Files containing valuable intellectual property (designs or formulations) should be identified and access to them logged.

Banking and financial systems are particularly prone to fraud, just because the benefits to the perpetrator are so direct. There are potential conflicts between the needs of confidentiality, management control, and auditing, and they must often be handled by using a common reference identifying each transaction.

Financial trading systems are increasingly electronic. There is a need for very high availability on the outward (information dissemination) path, and for confidentiality and authentication on the return (dealing) path. Most trading takes place within closed systems, and some stock exchanges are considering setting up certification authority systems.

Financial market information is regarded as a valuable commodity in its own right, and systems for its distribution either use private circuits or encrypted broadcast to protect their rights and their revenue. Other communications between banks and their customers use smart cards or tokens for authentication.

The payment card clearing system depends greatly on data communications. Although many of the data communications methods used would nowadays be regarded as rather insecure, the card itself is normally regarded as an easier target by criminals. France has already moved (and other countries are moving) to the use of smart cards for payment cards, which allow PIN checks and other cryptographic functions to be carried out on the card.

Subscription services face a number of additional problems, not least the confidence and contractual relationship with their customers should a security breach occur. Other issues include the means of making payment over a network, protection of one user from another, control of email and subscription control for broadcast services.

Control of copyright and other intellectual property rights is a major problem for many subscription networks, as well as for software distributors and administrators of large networks. When workstations can load or copy software, it is often necessary to hold and maintain a software audit to ensure that only legitimate software is being used.

The highest levels of security are demanded in military, police, and national security applications. Here all the tools are available to make networks very secure, but users must update their technology constantly in order to stay comfortably ahead of the code breakers. In this situation, the motives and resources available to break any security system can be very substantial. Such systems are beyond the scope of this book.

17.13   CONCLUSIONS

Data communications security is a many-sided problem. In addition to the technology, users must be aware of the environment, motives, cost and time factors, and the economics and working practices of the sectors they are involved in. They must know what problems are likely to arise and the potential cost to the company.

Data communications technologies are increasingly reliable. At the lowest level it is now possible to transfer data across huge distances with extremely low error rates. When cables and fiber optics are not available, radio, microwave, and satellite communications paths are readily available.

Differences in standards and applications mean, though, that it is not always so easy to send a message or a file to another user on another network. Supernetworks are growing up to meet more and more of these applications, but their security is much lower than that of other private or commercial VANs.

Cryptographic technology can meet most of today's requirements for full security through the use of data encryption, MACs, and digital signatures. Hardware tools such as smart cards are particularly powerful and should be considered by anyone responsible for a security-conscious network or database.

Password control at the beginning of a session or on entry to an application is another critical area. Again, tokens, smart cards, or other hardware devices are the best defense, but there are also software tools that can enhance security. Unfortunately, few of these tools are built into standard applications, and so the security must be added around the outside. Most mainframe operating systems provide adequate security services, but few PC or small systems can offer the same level of protection for applications or data.

Access rights management is the tool through which most confidentiality and interuser protection issues are addressed. The services offered by many operating systems are much too coarse or simply inadequate for the very powerful applications and valuable data requiring protection. Additional software packages are available to supplement these services.

No matter how good the technology used is, the security of a company's data depends as much on physical and human factors as on any hardware or software. Buildings, computer rooms, and telecommunications equipment must be well-designed and secured. It is best if anything that might provide temptation is kept out of sight. Electrical supplies and cabling systems deserve as much care as the logical design of the network.

Staff must be aware of the security implications of their actions and systems. Good procedures are often the best controls. An atmosphere of mistrust is likely to breed untrustworthy behavior. A crime normally has a motive, and knowledge and anticipation of these motives can dramatically reduce a company's susceptibility to all forms of fraud or security breaches, whether or not they are computer-related.

Most of these factors may simply be regarded as good practice, but the sheer range of the factors to be taken into account may be unfamiliar to many of those who take responsibility for the security of a data communications network. In these cases, as in so many others, the active support and good example of senior management will often be a deciding factor.

Glossary

ANSI	American National Standards Organization
ASIC	application-specific integrated circuit
ATM	asynchronous transfer mode
ATM	automated teller machine
bps	bits per second
CAD	computer-aided design
CAM	card authentication method
CB	Groupement Cartes Bancaires
CD-ROM	compact disk read-only memory
CALS	computer-aided acquisition and logistic support
CEN	*Centre Européen de Normalisation* (European Standards Center)
CLI	calling line identification
CMIP	Common Management Information Protocol
CRC	cyclic redundancy check
CSMA/CD	carrier sense multiple access with collision detection
CVM	cardholder verification method
DEA	Data Encryption Algorithm (same as DES)
DES	Data Encryption Standard
DSP	digital signal processor
EDI	electronic data interchange
EEPROM	electrically erasable read-only memory
E²PROM	electrically erasable read-only memory
EMI	electromagnetic interference
EDIF	Electronic Design Interchange Format
EDIFACT	Electronic Data Interchange for Administration, Commerce and Transport (standard)

EFT	electronic funds transfer
EFT-POS	electronic funds transfer at point of sale
EPOS	electronic point of sale (terminal)
ETSI	European Telecommunication Standards Institute
FDDI	fiber distributed data interface
FEAL	Fast Encryption Algorithm
FEP	front-end processor
GSM	Global System for Mobile Communication
IDEA	International Data Encryption Algorithm
IEC	International Electrotechnical Commission
IEEE	Institute of Electrical and Electronics Engineers
IP	information provider
ISDN	Integrated Services Digital Network
ISO	International Standards Organization
ITSEC	IT Security Evaluation and Certification Scheme
ITU	International Telecommunications Union (the successor to the CCITT)
kbps	kilobits per second
LAN	local-area network
LCD	liquid crystal display
LEO	low-earth-orbit (satellite)
MAC	message authentication check
MAN	metropolitan-area network
Mbps	megabits per second
MTBI	mean time between incidents
NCSC	National Computer Security Center
OSI	Open Systems Interconnection
PAD	packet assembler-disassembler
PC	personal computer
PCMCIA	Personal Computer Memory Card International Association
PEM	privacy-enhanced mail
PGP	Pretty Good Privacy
PIN	personal identification number
PMR	private mobile radio
PROM	programmable read-only memory
PSTN	public switched telephone network
PTT	post, telephone, and telegraph
RACF	resource access control facility
RAM	random-access memory
ROM	read-only memory
RSA	Rivest-Shamir-Adleman (public key encryption scheme)
SNA	System Network Architecture
SNMP	Simple Network Management Protocol

STEP	Standard for Exchange of Product Model Data
SSADM	Structured System Analysis and Design
SWIFT	Society for Worldwide International Funds Transfer
TCP/IP	Transmission Control Protocol/Internet Protocol
TCSEC	Trusted Computer System Evaluation Criteria
VAN	value-added network
VBI	vertical blanking interval
VLF	very low frequency
VSAT	very-small-aperture terminal (satellite terminal)
WAN	wide-area network

Bibliography

Standards

ANSIX3.92, "Data Encryption Algorithm." *American National Standards Organization.*

"Computer Security Subsystem Interpretation of the TCSEC: NCSC-TG-009." *NCSC*, New York, September 1988.

Information Technology Security Evaluation Criteria (ITSEC), "Provisional Harmonised Criteria." *Commission of the European Communities*, June 1991.

"Integrated Circuit Card Specifications for Payment Systems," *Europay International SA, Master-Card International. Inc., Visa International Service Association*, October 1994.

ISO 7498, "Open Systems Interconnection—Basic Reference Model." *International Standards Organization.*

ISO 7498-2, "Open Systems Interconnection Reference Model—Security Architecture." *International Standards Organization.*

ISO 8732, "Key Management to Achieve Security for Financial Institutions Engaged in Financial Transactions (Wholesale)." *International Standards Organization*, 1988.

ISO 9654, "Banking—Personal Identification Number Management and Security." *International Standards Organization*, 1991.

ISO/IEC 7811-7813, "Identification Cards—Financial Transaction Cards." *International Standards Organization*, 1990.

ISO/IEC 7816, "Identification Cards—Integrated Circuit Cards With Contacts." *International Standards Organization*, 1989.

ISO/IEC 8730, "Banking—Requirements for Message Authentication." *International Standards Organization*, 1990.

ITU Recommendation X.400, "Message Handling—System and Service Overview." *ITU*, Geneva.

ITU Recommendation X.435, "Message Handling Systems—Electronic Data Interchange." *ITU*, Geneva.

Trusted Computer System Evaluation Criteria ("The Orange Book"): DOD-5200.28-STD. *US Department of Defense*, December 1985.

Surveys

IT Security Breaches Survey 1994. National Computing Centre (UK), 1994.

Further Reading in Cryptography

Purser, M., *Secure Data Networking.* Norwood, MA: Artech House, 1993.

Rhee, M.Y., *Cryptography and Secure Communications.* McGraw Hill, 1994.

Simmonds (Ed.), *Secure Communications and Asymmetric Cryptosystems.* American Association
for the Advancement of Science, 1982.
Torrieri, D., *Principles of Secure Communication Systems.* Dedham, MA: Artech House, 1986.

Risk Management

Ansell, and Wharton (Eds.), *Risk: Analysis, Assessment and Management.* Wiley, 1992.
Ardis, *Risk Management: Computers, Fraud & Insurance.* McGraw Hill, 1987.
Ritchie, and Marshall, *Business Risk Management.* Chapman & Hall, 1993.

Technical EDI

Hendry, M., *Implementing EDI.* Norwood, MA: Artech House, 1993.

SGML

ISO 8879, "Specification for Standard Generalized Markup Language (SGML) for Text and
Office Systems." *International Standards Organization.*

STEP

ISO 10303, "Standard for the Exchange of Product Model Data." *International Standards
Organization.*

ODA

ISO 8613, "Office Document Architecture (ODA) and Interchange Format for Text and Office
Systems." *International Standards Organization,* 1990.

EDIF

ANSI/EIA RS 548, "EDIF Electronic Design Interchange Format." *American National Standards
Organization,* 1988.
Stanford, Mancuso (Eds.), *Introduction to EDIF.* Washington, DC: Electronic Industries Associa
tion, 1988.

About the Author

Mike Hendry has a degree in engineering from Cambridge and a degree in business administration from IMI Geneva. He is multilingual and has worked in engineering, marketing, project management, and consulting in almost every European country. Since 1982, he has worked as a freelance consultant, specializing in data communications and payment systems.

Index

The Artech House Telecommunications Library

Vinton G. Cerf, Series Editor

Introduction to Telephones and Telephone Systems, Second Edition,
A. Michael Noll

Introduction to X.400, Cemil Betanov

Land-Mobile Radio System Engineering, Garry C. Hess

LAN/WAN Optimization Techniques, Harrell Van Norman

LANs to WANs: Network Management in the 1990s, Nathan J. Muller and
Robert P. Davidson

Long Distance Services: A Buyer's Guide, Daniel D. Briere

Measurement of Optical Fibers and Devices, G. Cancellieri and U. Ravaioli

Meteor Burst Communication, Jacob Z. Schanker

*Minimum Risk Strategy for Acquiring Communications Equipment and
Services,* Nathan J. Muller

*Mobile Communications in the U.S. and Europe: Regulation, Technology, and
Markets,* Michael Paetsch

Mobile Information Systems, John Walker

Narrowband Land-Mobile Radio Networks, Jean-Paul Linnartz

Networking Strategies for Information Technology, Bruce Elbert

Numerical Analysis of Linear Networks and Systems, Hermann Kremer *et al.*

Optimization of Digital Transmission Systems, K. Trondle and Gunter Soder

Packet Switching Evolution from Narrowband to Broadband ISDN, M. Smouts

Packet Video: Modeling and Signal Processing, Naohisa Ohta

Personal Communication Systems and Technologies, John Gardiner and
Barry West, editors

The PP and QUIPU Implementation of X.400 and X.500, Stephen Kille

Practical Computer Network Security, Mike Hendry

Principles of Secure Communication Systems, Second Edition, Don J. Torrieri

Principles of Signaling for Cell Relay and Frame Relay, Daniel Minoli and
George Dobrowski

Principles of Signals and Systems: Deterministic Signals, B. Picinbono

Private Telecommunication Networks, Bruce Elbert

Radio-Relay Systems, Anton A. Huurdeman

Radiodetermination Satellite Services and Standards, Martin Rothblatt

Residential Fiber Optic Networks: An Engineering and Economic Analysis,
David Reed

Secure Data Networking, Michael Purser

Service Management in Computing and Telecommunications, Richard Hallows

*Setting Global Telecommunication Standards: The Stakes, The Players, and
The Process,* Gerd Wallenstein

Smart Cards, José Manuel Otón and José Luis Zoreda

For further information on these and other Artech House titles, contact:

Artech House
685 Canton Street
Norwood, MA 02062
617-769-9750
Fax: 617-769-6334
Telex: 951-659
email: artech@world.std.com

Artech House
Portland House, Stag Place
London SW1E 5XA England
+44 (0) 171-973-8077
Fax: +44 (0) 171-630-0166
Telex: 951-659
email: bookco@artech.demon.co.uk